Food, Senses and the City

This work explores diverse cultural understandings of food practices in cities through the senses, drawing on case studies in the Americas, Asia, Australia, and Europe.

The volume includes the senses within the popular field of urban food studies to explore new understandings of how people live in cities and how we can understand cities through food. It reveals how the senses can provide unique insight into how the city and its dwellers are being reshaped and understood. Recognising cities as diverse and dynamic places, the book provides a wide range of case studies from food production to preparation and mediatisation through to consumption. These relationships are interrogated through themes of belonging and homemaking to discuss how food, memory, and materiality connect and disrupt past, present, and future imaginaries. As cities become larger, busier, and more crowded, this volume contributes to actual and potential ways that the senses can generate new understandings of how people live together in cities.

This book will be of great interest to students and scholars of critical food studies, urban studies, and socio-cultural anthropology.

Ferne Edwards is a Postdoctoral Fellow in Socially and Environmentally Just Transitions, Department of Design, Norwegian University of Science and Technology, Trondheim, Norway, and was previously Research Fellow, RMIT University Centre for Urban Research, Melbourne, Australia, and Work Package Lead of the European Union's Horizon 2020 EdiCitNet project at RMIT Europe, Barcelona, Spain. Ferne is a cultural anthropologist researching edible cities, food waste, urban beekeeping, non-monetary food economies, and food sharing.

Roos Gerritsen is an anthropologist who works in social innovation and design. In her work she tries to bring science outside its academic bubble. She also works for an organisation that enables exchange through cooking. She worked previously at Heidelberg University and holds a PhD in cultural anthropology and development sociology from Leiden University, the Netherlands. Roos is the author of *Intimate Visualities and the Politics of Fandom in India* (2019).

Grit Wesser is a social anthropologist currently working on the AHRC-funded collaborative research project 'Knowing the Secret Police: Secrecy and Knowledge in East German Society' (2018–2021) at Newcastle University, UK. Previously, she taught social anthropology at the University of Edinburgh, UK, where she also earned her PhD in social anthropology (2016).

Routledge Studies in Food, Society and the Environment

Localizing Global Food
Short Food Supply Chains as Responses to Agri-Food System Challenges
Edited by Sophia Skordili and Agni Kalfagianni

Seafood Supply Chains
Governance, Power and Regulation
Miriam Greenwood

Civil Society and Social Movements in Food System Governance
Edited by Peter Andrée, Jill K. Clark, Charles Z. Levkoe and Kristen Lowitt

Voice and Participation in Global Food Politics
Alana Mann

Plant-Based Diets for Succulence and Sustainability
Edited by Kathleen May Kevany

Sustainable Food System Assessment
Lessons from Global Practice
Edited by Alison Blay-Palmer, Damien Conaré, Ken Meter, Amanda Di Battista and Carla Johnston

Raw Veganism
The Philosophy of the Human Diet
Carlo Alvaro

The Bioeconomy Approach
Constraints and Opportunities for Sustainable Development
Udaya Sekhar Nagothu

Resourcing an Agroecological Urbanism
Political, Transformational and Territorial Dimensions
Edited by Chiara Tornaghi and Michiel Dehaene

Food, Senses and the City
Edited by Ferne Edwards, Roos Gerritsen and Grit Wesser

For more information about this series, please visit: www.routledge.com/books/series/RSFSE/

Food, Senses and the City

Edited by Ferne Edwards, Roos Gerritsen and Grit Wesser

Routledge
Taylor & Francis Group
LONDON AND NEW YORK

earthscan
from Routledge

First published 2021
by Routledge
2 Park Square, Milton Park, Abingdon, Oxon OX14 4RN

and by Routledge
52 Vanderbilt Avenue, New York, NY 10017

Routledge is an imprint of the Taylor & Francis Group, an informa business

British Library Cataloguing-in-Publication Data
A catalogue record for this book is available from the British Library

Library of Congress Cataloging-in-Publication Data
Names: Edwards, Ferne, editor. | Gerritsen, Roos, editor. | Wesser, Grit, editor.
Title: Food, senses and the city / edited by Ferne Edwards, Roos Gerritsen, and Grit Wesser.
Description: Abingdon, Oxon ; New York, NY : Routledge, 2021. | Series: Routledge studies in food, society and the environment | Includes bibliographical references and index.
Identifiers: LCCN 2020044085 (print) | LCCN 2020044086 (ebook)
Subjects: LCSH: Food consumption. | Food habits. | Nutritional anthropology. | City dwellers. | Cities and towns—Social aspects.
Classification: LCC GN407 .F685 2021 (print) | LCC GN407 (ebook) | DDC 394.1/2—dc23
LC record available at https://lccn.loc.gov/2020044085
LC ebook record available at https://lccn.loc.gov/2020044086

ISBN: 978-0-367-45823-2 (hbk)
ISBN: 978-0-367-72362-0 (pbk)
ISBN: 978-1-003-02558-0 (ebk)

Typeset in Bembo
by Apex CoVantage, LLC

Contents

Preface

This book emerged from a panel at the fifteenth Biennial Conference of the European Association of Social Anthropologists (EASA) held in Stockholm (EASA 2018) to explore different contextual understandings and methodologies that bring together the senses and food in the city. The theme of the conference 'Staying, Moving, Settling' recognised the dramatic mobility characteristic of modern society to consider both spatial movement alongside the 'backgrounds, forms and contexts, and longer-term implications' of mobility in general (EASA 2018: 9). 'Staying' is described in the conference programme as 'still the normal way of life, "business as usual"' (ibid) to acknowledge that choices are at times enforced with consequences for those left behind. 'Moving' connotes varied forms of departure – when it may be planned or in crisis, alone or with others, in one swift go or in stages of transition. 'Settling', then, considers what happens when one arrives – the relationships between natives and newcomers; processes of integration that may involve (or not) the creation of new networks, skills, and identities; and what you hold dear from the life left behind. This conference theme well-complemented our interests to expand on literature in critical food and urban studies.

The next step in this book's journey involved participating in the fourteenth Congress of the International Society for Ethnology and Folklore held in Santiago de Compostela, Spain (14–17 April 2019). This conference's theme, 'Track Changes: Reflecting on a Transforming World', dealt with the 'processes and practices of transformation – as ways of being and as processes of becoming' (SIEF 2019). 'Track Changes' called for contributors to 'track the changes that take place at different scales, speeds and intensities', where conference organisers sought 'to encourage researchers to follow something that moves or alters by noticing the marks or signs that it has left behind' (SIEF 2019). 'Track Changes' recognises both where knowledge has been and where it is going, in addition to acknowledging the process of knowledge acquisition through reflection. In this resulting volume, contributors build on work in anthropology and associated social science disciplines to expand territories of 'food, senses, and the city', based on their experiences of changing worlds.

At SIEF, the editors also experimented with multi-sensory methodologies by hosting a sensory food walk workshop. The food walk allowed us to explore the streets and surroundings of Santiago de Compostela, a city resplendent in its own rich food history. Recognising that we were external to this region, we relied on our senses influenced by our immediate surrounding environment to direct and support our interpretations. Branching into three groups, one group tasted local delicacies from the region, accompanied by tales of their folklore. A shop window boasted local fishing resources, where a participant from the region described their tastes and local values and recounted how the seafood was caught, eaten, and prepared. Walking through the streets, we were also struck by its longevity as we saw weeds breaking through the stone walls and cobbled streets while all around us, we were bombarded by the chatter of pilgrims on their way. These two key events, coupled with the helpful comments from the book proposal reviewers and our chapter contributors, have shaped and (re)shaped this book's form, purpose, and structure, to consider new pathways for 'food, senses, and the city'.

References

European Association of Social Anthropologists (EASA) (2018) 'easa 2018 Staying, Moving, Settling', Conference Programme, 5th Biennial Conference of the European Association of Social Anthropologists, Stockholm, Sweden, 14–17 August. DOI: 10.22582/easa2018prg.

SIEF (2019) 'Track Changes: Reflecting on a Transforming World', 14th Congress of Société Internationale d'Ethnologie et de Folklore (SIEF), Santiago de Compostela, Spain, 14–17 April. https://www.siefhome.org/congresses/sief2019/theme.shtml.

Acknowledgements

Shuhua Chen would like to express her debt of gratitude to Yang Cui, who was generous in sharing her life and stories. Special thanks to Professor Walter Hakala, Professor Nigel Rapport, Dr Stephanie Bunn, Dr Grit Wesser, Dr Roos Gerritsen, and Dr Ferne Edwards for their constructive feedback for the chapter. The research was made possible thanks to funding from the Centre for Cosmopolitan Studies, University of St Andrews, and the Foundation for Urban and Regional Studies.

Carole Counihan warmly thanks Professors Gabriella Da Re and Benedetto Meloni and the Visiting Professor and Visiting Scholar programs for invitations to the University of Cagliari in 2011, 2015, and 2016 and Professor Franco Lai for an invitation to the University of Sassari in 2014. Thanks to my husband, anthropologist Jim Taggart, for participating in the fieldwork with me and offering many thoughtful insights.

Ferne Edwards would like to thank Associate Professor Jane Dixon for her guidance and support and the Commonwealth Scientific and Industrial Research Organisation for the opportunity to begin research in this field as part of the Urbanism, Climate Adaptation and Health program. While this chapter draws on independent research conducted since that period, the initial momentum and encouragement are greatly appreciated.

Aitzpea Leizaola would like to thank the peer reviewers for their insightful feedback. Part of the present work owes much to a previous project I coordinated on the evolution of Basque gastronomy at the eve of the millennium that received funding from the Basque government in 2009–2010. I would like to thank my family: Miren Egaña, Fermin Leizaola, Usue Leizaola, and especially Arkaitz Garmendia. As childcare was closed for several months during the COVID-19 confinement, their support has been essential to take care of our toddler son while I was writing.

Daniel Monterescu acknowledges the generous fieldwork funding by the Central European University Research Support Scheme for Chapter 6.

Paz Saavedra, José David Gómez-Urrego, and J. Guillermo Gómez-Urrego's deepest gratitude goes to Luisa and Alicia for opening their home and sharing their stories. We hope to bring more attention and care to their invaluable work with their chapter (pseudonyms provided). Paz Saavedra gratefully acknowledges that this chapter was informed by her doctoral dissertation on traditional practices of care in Ecuador funded by SENESCYT under the International Scholarship Program. She would also like to thank SEDAL, the institution in charge of implementing Bio-Vida, and particularly Patricia Yacelga, the director, who gave her access to the *feria* in the first place and kindly shared its history. Furthermore, the authors want to acknowledge the crucial collaboration of the people forming Bio-Vida, whose stories inspired the reflections put together in this chapter. Also, the authors would like to thank Paul B. Vallejo for his crucial collaboration during the field part of the research. We would also like to acknowledge the help and support of Dr Ferne Edwards in the writing of this chapter; her edits and feedback were very important to the final product.

Monica Stroe would like to acknowledge that Chapter 12 was written as part of the project 'Material projects of class distinction: An analysis of middle classness in postsocialist Romania from a material culture perspective', funded by the Romanian Executive Agency for Higher Education, Research, Development and Innovation Funding Programme 'Human Resources', PN-II-RU-TE-2014-4-2650.

Vincent Walstra gratefully acknowledges that Chapter 2 draws on research that received funding from the European Research Council (ERC) under the European Union's Horizon 2020 research and innovation programme (grant agreement No 724151).

Grit Wesser would like to thank fellow cake lovers Imogen Bevan, Janet Carsten, and Jan Dobbernack as well as Roos Gerritsen, Ferne Edwards, and the peer reviewers for their constructive feedback on earlier versions of Chapter 7. Special thanks go to the Thuringians who allowed her to participate in their life-cycle celebrations and who ate and talked cake with her.

Premila van Ommen would like to thank food dream producers Natalia Munatajeva, Ojesh Singh, Adelino Santos, and the TECHNE Dumpling Crew. Her thanks to the participants in this research who allowed her to share their social media worlds. She also gratefully acknowledges that her chapter is informed by doctoral dissertation work funded by the Techne AHRC Doctoral Training Partnership.

Figures

Abbreviations

ASMR	Autonomous Sensory Meridian Response
EASA	European Association of Social Anthropologists
GAS	Gruppo d'Acquisto Solidale (Solidarity Purchase Group)
GDR	German Democratic Republic
GMA	Guadalajara Metropolitan Area
HCMC	Ho Chi Minh City
IGTV	Instagram TV
LGBT	Lesbian, Gay, Bisexual, and Transgender
NAFTA	North American Free Trade Agreement
SIEF	Société Internationale d'Ethnologie et de Folklore
UNESCO	United Nations Educational, Scientific, and Cultural Organisation

Contributors

Editors

Ferne Edwards is a Postdoctoral Fellow in Socially and Environmentally Just Transitions, Department of Design, Norwegian University of Science and Technology (NTNU), Trondheim, Norway, and was previously Research Fellow, RMIT University Centre for Urban Research, Melbourne, Australia, and Work Package Lead of the European Union's Horizon 2020 EdiCitNet project at RMIT Europe, Barcelona, Spain. Ferne is a cultural anthropologist researching edible cities, food waste, urban beekeeping, non-monetary food economies, and food sharing.

Roos Gerritsen has worked for over fifteen years as an anthropologist at Leiden University and Heidelberg University. Her work on Tamil fan clubs and visuality has been published with AUP (2019) as *Intimate Visualities and the Politics of Fandom in India*. She has also published in journals such as *Ethnos* and *Visual Anthropology*. Currently, Roos works for Über den Tellerrand. She is active in the fields of smart city, food, and migration in different capacities and is involved in different projects at the interface of science communication, art, and social innovation.

Grit Wesser is a social anthropologist and currently works on the AHRC-funded collaborative research project 'Knowing the Secret Police: Secrecy and Knowledge in East German Society' (2018–2021) at Newcastle University, UK. Previously, she taught social anthropology at the University of Edinburgh, UK, where she also earned her PhD in social anthropology (2016). Besides the anthropology of food, her research interests include East Germany, memory and history, Cold War studies, social change, kinship and gender, feminisms, the anthropology of the state, and ritual and personhood.

Contributors

Diti Bhattacharya is a resident adjunct with the Griffith Centre for Social and Cultural Research at Griffith University. Her area of research expertise includes human and cultural geography, migration and mobilities, tourism geographies, and critical heritage and museum geographies. Her doctoral thesis examined spatial movements and material attachments in the second-hand book market of College Street, Calcutta. She combines her research practice working as a research assistant, sessional lecturer, and tutor and as a freelance writer and photographer for various publications.

Melissa S. Biggs is a cultural anthropologist specialising in issues of representation and cultural heritage, with a focus on food and museums. From 2016 to 2017, she was a Fulbright Garcia-Robles Scholar located in Guadalajara, Jalisco, where she carried out research on culinary tourism and traditional cooks. She is currently the guest curator for the international participatory exhibit 'Hostile Terrain 94' at its Austin, Texas, site. Previous projects include 'Native American Gaming and Self-Representation', which examined the relationships between Native casinos, museums, and cultural centres, and her dissertation, 'Exhibiting Mexicanidad: The National Museum of Anthropology and Mexico City in the Mexican Imaginary'.

Shuhua Chen is a Research Fellow at the Department of Social Anthropology, Norwegian University of Science and Technology. She was a Research Fellow at the University of Birmingham and a Visiting Scholar at the University at Buffalo after she received her PhD in social anthropology at the University of St Andrews in 2018.

Carole Counihan is Professor Emerita of Anthropology at Millersville University and has been studying food, gender, and culture in Italy and the USA for 40 years. She is the author of *Italian Food Activism in Urban Sardinia* (2019), *A Tortilla Is Like Life* (2009), *Around the Tuscan Table* (2004), and *The Anthropology of Food and Body* (1999). She is co-editor of *Food and Culture: A Reader* (1997, 2008, 2013, 2018), *Taking Food Public* (2012), *Food Activism* (2014), and *Making Taste Public* (2018) and is editor-in-chief of the scholarly journal *Food and Foodways*.

Catherine Earl is Lecturer in Communication at RMIT Vietnam. Author of *Vietnam's New Middle Classes: Gender, Career, City* (2014) and editor of *Mythbusting Vietnam: Facts, Fictions, Fantasies* (2018), Catherine has published extensively on the changing nature of work and welfare, migration, and gender and social change in contemporary Vietnam and Australia. Her current research focuses on digital use among middle and affluent classes in the mega-urban region of Ho Chi Minh City.

José David Gómez-Urrego has a PhD in science and technology studies from the University of Edinburgh. He received training as a sociologist in the Pontificia Universidad Católica del Ecuador (PUCE) and studied his MSc and PhD in the Centre for Science, Technology and Innovation Studies (STIS) at the University of Edinburgh. He is interested in the intersections between STS and social theory around time, particularly about the multiple roles temporalities have in collective practices, sociotechnical regimes, and cognition.

J. Guillermo Gómez-Urrego is a sociologist, specialising in community-based critical action research, with a focus on popular education and public health. An agroecology apprentice, he participates in and contributes to the construction of inclusive agroecological networks that incorporate social, cultural, economic, environmental, and political dimensions. He has an MSe in critical studies in education and an MSa in international development from Ohio University. He has a special interest in designing and implementing critical and holistic social projects and has extensive experience in horizontal and democratic work with grassroots communities.

Joel R. Hart is a DPhil candidate in anthropology at the University of Oxford. His doctoral research was conducted in two ethnically diverse low-income neighbourhoods on the Southern borders of the mixed Jewish-Palestinian city of Jaffa, Israel/Palestine. His ethnography intertwines materiality and sociality to illustrate how the relationship between the state and peripheral urbanism delimits the potential of Israeli multiculturalism. With a strong interest in the anthropology of food, he has also conducted research on the politics of culinary heritage in the Middle East.

Aitzpea Leizaola is Associate Professor of social anthropology and director of the anthropology master's program at the University of the Basque Country (UPV/EHU). Her research interests are mainly in political and symbolic anthropology, with a special focus on border studies, identity, and heritage, including food, memory studies, and popular forms of protest. She directed *Ahoy, pirates!* (2013), an ethnographic film on the transformation of the summer fiestas in Donostia (Visual Fest prize 2014, Rome). She has carried out extensive and multi-sited fieldwork in the Basque Country, Spain, and Turkey.

Daniel Monterescu is an Associate Professor of anthropology at the Central European University, Budapest. He holds a PhD from the University of Chicago and a Sommelier certificate in Italy and is training for completion of the Wine & Spirit Education Trust diploma in Austria. His current research focuses on gastronationalism and border wines in Europe and the Middle East through the concepts of terroir and territory. He is the author of *Jaffa Shared and Shattered: Contrived Coexistence in Israel/Palestine* (Indiana UP 2015) and coauthor of *Twilight Nationalism: Politics of Existence at Life's Edge* (Stanford UP 2018). His article 'Liquid Indigeneity: Wine, Science and Colonial Politics in Israel/Palestine' was published by *American Ethnologist* (with Ariel Handel, 2019).

Paz Saavedra is a sociologist researching the intersecting politics of care, knowledge, and time. She has a PhD in interdisciplinary studies from the University of Warwick. Her research follows the reproduction of situated, embodied, and intergenerational knowledge in more-than-human worlds. Particularly, her work focuses on the politics of knowledge production in traditional practices of care and healing. She also works with community-based projects applying interdisciplinary methodologies in the development of pedagogical tools to generate communities of learning that weave together different local knowledges and practices.

Monica Stroe teaches anthropology at the Department of Sociology of the National University of Political Studies and Public Administration in Bucharest (Romania). Her current research interests include food and social class, morality and consumption, consumption and the senses, food entrepreneurship, food and microbiopolitics, and food heritage. She recently published *Preserves exiting socialism: Authenticity, anti-standardization and middle class consumption in postsocialist Romania*, in Evgenia Krasteva-Blagoeva (ed) *Approaching Consumer Culture: Global Flows and Local Contexts*, Springer (2018). Her latest research project looks at domestic sourdough breadmaking during the COVID-19 pandemic.

Premila van Ommen is a TECHNE-funded PhD candidate in cultural studies at the London College of Fashion, University of Arts London. Her research focuses on the impact of military Gurkha heritage on the cultural production and creative industries of young Nepali men in the UK. She is also the founder of the online photo archival project Urban Arhats, the Himalayan food collective Yak Bites, and the Afro-Nepali food/arts movement MOMOLIFE.

Vincent Walstra is a PhD member of the ERC-research team Food Citizens led by Professor Cristina Grasseni at the Institute of Cultural Anthropology and Development Sociology at Leiden University. He graduated with a research on urban agriculture in Utrecht and has continued studying the anthropology of food in the Netherlands, researching the nexus of food procurement and citizenship in Rotterdam.

1 The 'food, senses, and the city' nexus

Ferne Edwards, Roos Gerritsen, and Grit Wesser

Food enters, moves through, settles into, disrupts, and redesigns cities in novel ways: as community, school and allotment gardens, and foraging sites; as health food stores and farmers' markets; as freegan and vegan forms of protest and dietary reformation; as social treatise at the shared table; and by passing through as food trucks and as new forms of food delivery. Food may be grown, processed, cooked, consumed, and shared. Each engagement produces tactile, affective, visceral, and embodied relationships between people, places, and products that can instigate and uphold social relationships whilst embodying shifting values, meanings, and politics. Peoples' engagement with food in turn influences the shape and feel of the city, fostering the potential to bring people either together or apart, to connect or repel people from having a connection to place. Acknowledging the senses through urban food practices thus serves as an essential means by which to both link people to each other and to where they live.

Through our senses, we make sense of ourselves and the world. A palpable moment of nostalgia evoked by hearing a song from our youth or by encountering a smell that recalls a place where you used to hear the *chai wallah* (tea seller) from afar, revealing their presence and almost making you smell the sticky, milky tea that they sell. The smell of the pizza restaurant downstairs irritates us, but as soon as the new owner gives us a pizza occasionally, the greasy aroma lingering in the air is suddenly less disturbing. Think of walking through a city to absorb its atmospheres of colourful murals; of the passing by of tourists slurping iced chai lattes in the summer heat; of tasting fresh produce on the tip of a wooden spoon at a farmers' market; of ordering a treat from a food truck or a drink from a street bar; of eating in the city's darkness, perhaps in one of its nearby parks; or of dining in the brightness in one of its fancy Michelin star restaurants.

Such examples already indicate that how we sense and make sense of the world around us is not merely an individual but also a socio-cultural act (Howes and Classen 2014). In this book, we position ourselves between a phenomenological and a cultural approach to the study of the senses. We do not see experience as merely embodied, nor do we see the senses as solely a cultural construct. We are not studying the senses per se, but we are *consciously studying with the senses*, allowing a focus on the senses to give us a deeper understanding of the food, city, and the senses nexus. Moreover, the senses can be both an object of study and a means of inquiry (Howes 2019: 18), and in this work, we are mostly concerned with the latter.

This approach raises multifarious questions. What role do the senses play in the production, preparation, and consumption of food? How do urban food practices conjure up memories of home for new arrivals or provide a means for understanding those who remain? How are social relations and distinctions reproduced and reshaped through introduced and diverse preparation styles? How do people embody and remember transformations – social, economic, cultural, historical, political – and their materialisation in their everyday lives? Does sensing food materialise vulnerability, uncertainty, the unreliable, the risky, the fragile, or the improvised?

In this volume, we explore how the study of the senses can provide a more holistic, thick description of urban experiences. A sensory ethnography is not necessarily an ethnography that investigates how the senses are used but rather looks at the ways in which sensory experiencing and knowing make sense of people's everyday lives (Pink 2009). We examine everyday life and the various contexts in which culturally shaped sensory properties and sensory experiences of food are invested with meaning, emotion, memory, and value (Sutton 2010).

This volume expands the increasingly popular field of urban food studies to include the senses; we explore understandings of how people live in cities and how we can understand cities through food. It brings together social science research grounded in rich ethnographic accounts from diverse urban centres around the world to ask how the city and food co-produce each other. Drawing mainly from anthropological accounts informed by related disciplines, this volume asks how the senses can provide unique insights into city life.

Food and its production, preparation, consumption, and mediatisation move through time and space, creating new forms of conviviality, commensality, and sociality. Diverse cultural interpretations, based on both uniting and separating forces of food practices, allow cities to be reconceptualised as 'many places within one', revealing new worlds of dynamic cultural engagements that can benefit richer understandings for present and future forms of urban sociabilities. These relationships are interrogated through themes of belonging and homemaking to discuss how food, memory, and materiality connect and disrupt past, present, and future imaginaries. As cities become larger, busier, and more crowded, this volume contributes to actual and potential ways that senses can generate new understandings of how people live together or create boundaries in cities. This new direction in both theory and practice extends beyond the dominant focus on larger Euro-American cities to include cities, places within cities, and references to the city in Central and South America, Australia, and Asia.

In this introduction, we explore the nexus of 'food, senses, and the city' in theory and practice. We start with a literature review of the 'sensory turn' in the social sciences, to acknowledge key debates and concepts in food and urban studies that, in turn, influence this volume's approach. David Howes (1991: 8) reminds us that there are many 'ways of sensing the world'; in order to capture the senses, new methodologies need to be developed. This book brings to the fore research methodologies that go beyond the written word applied through grounded case study material. Finally, this chapter summarises chapters in the sections; 'The city and its other', 'The past in the present: memory and food', and 'Disrupting and re-imagining'.

The sensory turn in the social sciences

A brief history

Howes's early work explores 'how the patterning of sense experience varies from one culture to the next in accordance with the meaning and emphasis attached to each of the modalities of perception' (1991: 3). This approach later shifted to a form of 'sensorial fieldwork' (Robben and Sluka 2007), in which an anthropologist's sharing the senses of a culture to make sense of it extended traditional interpretations of participant observation. This focus on sensation transformed into a new focus on interpretation introduced by Clifford Geertz (1973), to shift once more to a focus on representation in the 1980s with *Writing Culture: The Poetics and Politics of Ethnography* by James Clifford and George E. Marcus (1986). Marcus regrets that the later discussion focused largely on the textual and ethnographic authority; while he concentrated more on the sensorial and aesthetic concerns in his work with Calzadilla on a Venezuelan market (Calzadilla and Marcus in Cox *et al.* 2016). Other methodologies and sensorial explorations further aided this shift (Taylor 1996; Grimshaw 2007; MacDougall 2005), in which both sound (termed 'acoustemology' by Feld 1991; Feld and Brenneis 2004) and taste ('gustemology' per Sutton 2001, 2010) joined the visual in sensory research. Paul Stoller (1997) took the opportunity of the writing debate as a moment to argue for a 'sensuous scholarship' in which the researcher's embodied presence and modes of representation evoke a kind of sensuality instead of treating the senses as an object of study, an approach reflected in several chapters (see Battacharya, Edwards, Gerritsen, and Stroe, this volume). Cultural extensions of what in Western tradition is considered to be the senses were further expanded by Kathryn Geurts (2003), who explored 'attention' through recognising a range of indigenous senses. In her detailed study of the Anlo Ewe in Southeastern Ghana, Geurts goes beyond arguing that sensoria vary cross-culturally. Rather, she succeeds in demonstrating the significance of the Anlo Ewe sensorium – including 'balance' – for shaping every aspect of social life: moral codes, sense of place, socialisation, and personhood.

Since the 1980s, the senses in the social sciences have begun to receive considerable attention, aptly labelled 'the sensory turn' (Howes 2019). From first studying each of the five senses to later developing into an anthropology of the senses (Howes 1991), through to the meshwork of experience (Ingold 2008) and sensory anthropology as a way of conducting research (Pink 2009), ways to work with the senses vary greatly.

Key methodological debates in sensory anthropology

A key debate between anthropologists Sarah Pink, David Howes, and Tim Ingold in *Social Anthropology* lays bare questions of understanding sensory perception and how culture is understood through the senses (Pink and Howes 2010; Ingold 2011). This section describes three aspects of this debate: challenging an ontogenetic phenomenology, individual versus multi-sensorial perspectives, and singular versus trans-disciplinary approaches for studying the senses in the social sciences.

Beyond a phenomenological perspective

Ingold's work became the focus of criticism by Pink and Howes; Howes accused Ingold of staying within a limited phenomenological understanding of the sensual qualities of experience that tended 'to ignore how shared meanings shape the most "natural" of human actions and perceptions in dance and in life, slighting the cultural content inherently implied by physical and cultural experience [Bull 2018 (1997): p. 263]' (Howes 2019: 20). According to Howes, Ingold did not take into account the way in which perception is a cultural construct and left out some of the lower-ranked senses.

Ingold (2011), in turn, criticised Howes's approach, calling for a refocussing of sensory anthropology based on experience and perception drawing on Maurice Merleau-Ponty (1962) and James Jerome Gibson's ecological psychology (1979). Ingold suggests that separating out the senses, as proposed by Howes, situates them within a disembodied 'culture' that is incompatible with anthropology's stance on situated and embodied knowledge (Pink 2010). Furthermore, Ingold accused Howes of his own limited claims of seeing neuroscience within a historical and cultural paradigm, therefore undermining his own claim about indigenous sensory systems. In other words, by claiming that neuroscience is also part of a certain paradigm, one would actually not be able to use notions out of this paradigm to understand other paradigms. While we are not interested in taking a stance in this debate, we describe it at some length because we do see merit in its key points. In this book, we are neither merely following Howes in his cultural approach nor taking a purely phenomenological approach. Instead, we see a study of the senses in both, where individual experience is made by its environment, and this environment is made by socio-cultural and individual contexts and experiences. The ways in which the sensoria create and are created by experience, action, and context, and the ways in which individuals make places with and through the senses, are the common thread through the chapters in the book.

Towards a multi-sensory anthropology

Various authors have pointed to the limitations and simplification of thinking in five sensory modes as part of human experience and question the Western five-sense model (Vannini *et al.* 2012: 7). As Phillip Vannini *et al.* (2012: 7) express:

> modes of sensing inevitably blend and blur into one another, thus making their alleged boundaries fuzzy and indistinct in experience. It is this

> ecology of affective relations that should be the focus of our attention (see Howes 2003; Ingold 2000; Thrift 2008).

Instead of focusing on how the five senses individually create meaning in human lives, we follow Vannini *et al.* (2012) and acknowledge the engagement and fuzziness of sensorial experience. David Sutton (2010) refers to this union of the senses as 'synesthesia'. Moreover, synesthesia also 'blurs the objectivity and passivity of western sensory models by showing the ways that sensory experience is not simply passively registered but actively created between people' (Sutton 2010: 218).

As the editors experienced during a workshop in Santiago de Compostela (see Preface), to focus on a single sense omitted the presence and value of others. For example, walking through the cobbled-stone streets of Santiago de Compostela in search for food to eat together, for some the smell of freshly baked goods triggered happy childhood memories of bakeries that no longer exist, trumping the sound of our surroundings and the shape as well as the taste of the actual baked goods that differed from such memories. Some focused on the shape, texture, or colour of baked goods in choosing a particular item before they assessed it through comparing it with their own middle-class ventures into bread baking while others took photographs to share on their Instagrams. Such practices and articulations of varying preferences among a group of people highlight that taste (as social sense) is not simply an individual preference but a cultural practice that makes taste public (Højlund 2015; Counihan and Højlund 2018). Even separating such sensibilities immediately excludes the collective experience of tasting together, memories, and expectations, to name a few. The sense-making process is not merely physiological but entails minded and embodied social and cultural practices (Vannini *et al.* 2012: 15).

Sidney Mintz's work on sweetness and the socio-political history of sugar also illustrates this union across the senses and their transformative effect. He notes that a propensity for sweetness 'cannot possibly explain differing food systems, degrees of preference, and taxonomies of taste – anymore that the anatomy of the so-called organs of speech can "explain" any particular language' (Mintz 2005 [1985]: 113). Mintz's highly influential work on sugar and sweetness reveals how the distinctive history of sugar has changed not only eating practices and sensorial attachments but also the 'notions of time, gender, class, senses of self in relation to family, community and labor, and the "locus of desire" (Mintz 1996: p. 79)' (Sutton 2010: 212). This has been a crucial acknowledgement because it shows how the sensoria and subjectivities change throughout time and relate to significant socio-cultural notions.

An anthropological or interdisciplinary approach to studying the senses?

Pink (2010: 331) opened the aforementioned debate by outlining two possible strands: 'the original anthropology of the senses on the one hand, and the newer sensory anthropology on the other', in which a sensory anthropology 'both has its roots in and departs from the anthropological study of sensory perception and categories that characterises the anthropology of the senses' while sensory anthropology 'implies a "re-thought" anthropology, informed by theories of sensory perception, rather than a sub-discipline exclusively or empirically about the senses' (ibid). This newer vein of sensory anthropology presupposes an interdisciplinary turn in which

> a future for sensory anthropology that is embedded in a context of interdisciplinarity – in terms of the principles that inform a sensory approach, the 'sharing' of research methods across disciplines and the potential for interdisciplinary collaboration in the production and dissemination of research.
> (Pink 2010: 333)

Howes (Pink and Howes 2010) questions this departure on grounds of a sensory anthropology that has always been interdisciplinary, both a subject of study and a means of inquiry. This debate raises important disciplinary questions for this volume's authors: how do we study the multi-sensory environment, and how do we sense and make sense of it? How do we translate lived experience, and how do we represent it? What disciplines are we engaging with and drawing from to further develop methodological approaches?

We recognise that these are not only phenomenological questions but also questions of methodology. Much attention in exploring the senses and transforming urban environments has gone to visual change (Degen 2008: 9; see also Featherstone 2010; Ghertner 2011); in sensory ethnography, various scholars have used video and photography to attend to embodied understanding (MacDougall 2005; Pink 2009; Castaing-Taylor and Paravel 2012). More and more anthropologists have been working with the affordances of 'multimodal anthropologies', a term that has come to stand for the audiovisual and mediatic methodologies that also attend to the senses.

How to capture smell and represent it beyond writing? Or sound? The absence of smell and taste is partly due to the fact that no technologies such as the photograph or video camera exist for flavours and aromas (Rhys-Taylor 2017: 17). We admit that in this book we remain largely bound by text and images. A sensory analysis, we contend, is not a separate subfield of the study of the senses, but we follow Constance Classen in that sensory experiences are a fruitful perspective from which we can do ethnographic work and touch upon anthropological concerns (Classen 1997: 409).

Positioning the 'food, senses, and the city' nexus

Food

Food is a basic need that not only nourishes bodies and minds but also is a main ingredient of culture and social organisation. Because of its social significance, food already features in many anthropological classics – in descriptions of eating and drinking; food provisions, production and exchange; and ritual and symbolism (for example, see Evans-Pritchard 1940; Levi-Strauss 1964; Malinowski 1935; Mauss 1990 [1925]). Yet food and foodways only became legitimate main subjects of study in the late twentieth century (Watson and Klein 2016). James L. Watson and Jacob A. Klein (2016) ascribe the greatest impact on the establishment of the anthropology of food to two monographs: Sidney Mintz's (1985) *Sweetness and Power* and Jack Goody's (1982) *Cooking, Cuisine and Class*, noting that:

> Characteristic of both these works was their attempt to move beyond entrenched debates between proponents of 'symbolic' and 'materialist' theories, advocating historically and ethnographically grounded studies that explored the relationship between material practices, power and meaning. Further, by using patterns of food and drink production, distribution, consumption, communication and taste to explore the increasingly transnational connections shaping social relations and everyday experiences, Mintz and Goody demonstrated that the study of food was in fact a key way into the study of modern life.
>
> (Watson and Klein 2016: 3)

Each engagement with food – be it as grown, processed, cooked, consumed, or disposed of – produces tactile, affective, visceral, and embodied relationships that can instigate and uphold social relationships whilst embodying shifting values, meanings, and politics. Disposal, while recognised by Goody, has only begun to gain attention in recent years, when the 'problem' of food waste has moved up the global political agenda (Alexander *et al.* 2013; Evans *et al.* 2013).

In anthropological inquiry, the relationship between food and the senses can be traced back to structuralist keystone works by Claude Levi-Strauss (1983 [1964]) and Mary Douglas (1982), who both apply food as a tool 'oriented toward abstracting binary patterns in sensory features that reflect other structured aspects of "the food system" and its relationship to "the social system"' (Sutton 2010: 210). The sensorial qualities of food have been studied extensively with respect to social distinction, by key flavour principles and as central to exploring culture (Sutton 2010). Pierre Bourdieu (1984) is a key influence on the topic of taste; taste for Bourdieu (1984) is about distinction and class and does not in any way refer to the flavour, to the materiality of the food itself. Taste, here, becomes an aesthetic judgement and not a multi-sensory one 'which involves the dissolving of the object into the subject' (Borthwick in Sutton 2010: 211). Moreover, the senses are often seen in hegemonic relations, creating hierarchies, or understood as part of immigrant landscapes (Manalansan 2006; Law 2001).

Some chapters in our volume show evocatively how people embody notions of class and space through specific pasts and are thus sensed and shaped differently cross-culturally. For example, Monica Stroe (Chapter 12) shows how urban middle-class foodie flaneurs seek out *mici*, a traditional working-class food, to experience 'unhealthy working class' commensality as an eatertainment event at Bucharest's periphery. While such eatertainment brings into sharp relief Bucharest's gentrification (a process we can witness in cities worldwide), *mici* commensality, strongly associated with the socialist tradition of state-sponsored Labour Day celebrations, is sought, performed, and mocked by foodie flaneurs simultaneously (and, by extension, a particular socialist past that glorified the working class).

Food and the senses have continued to expand and diverge in application over the last decade. Here, we explore sub-themes that emerge throughout this book: sociality, commensality, conviviality, and memory in relation to food and the senses.

Sociality, commensality, and conviviality

The truism that humans are social beings – in cities or elsewhere, most of our lives become significant or meaningful through the interactions we have with others – is variously reflected through the conceptions of 'sociality', 'conviviality', and 'commensality'. Since they entail varying definitions and applications, in this section we position ourselves on how these concepts are used – explicitly or implicitly – in this book.

In recent decades, sociality has been increasingly employed because, unlike previous approaches to a more static inquiry into 'society' or those who view 'the social' as a product of social relations rather than being coextensive, its processual nature reflects that the social needs individual agency and that the two are constitutive of each other (Long and Moore 2014: 2). Nicholas J. Long and Henrietta Moore (2014) acknowledge that sociality's definition remains obscure, but they view this challenge less as an obstacle than as an indication of human sociality's capacity to take multifarious forms. Rather than using a narrow conception of sociality, they propose 'a theory of human sociality – which is to say any sociality involving humans – that can account for its diverse manifestations, its plasticity, and fragility, and also its possible resilience' (Long and Moore 2014: 2). Building upon the models proposed by Christina Toren and Marilyn Strathern in the 1989 Manchester Debate on Anthropological Theory (Strathern *et al.* 1990), Long and Moore thus understand sociality as 'a dynamic relational matrix within which subjects are constantly interacting in ways that are co-productive and continually plastic and malleable' (2014: 4). Crucially, sociality, as Toren notes, then focuses our attention on 'dynamic social processes in which any person is inevitably engaged rather than a set of rules or customs or structures or even meanings that exists *as a system* independently of the individual who is to be socialised' (Strathern *et al.* 1990: 19; emphasis in original).

Sociality shares this emphasis on dynamic social processes with conviviality. Akin to the German *Geselligkeit* in its English sense, conviviality connotes a festive, jolly – and often inebriated – togetherness or atmosphere, which is frequently facilitated through the consumption of food and particularly alcohol in the company of others (Overing and Passes 2000). The different ways conviviality is achieved – often in association with commensality – in various cross-cultural contexts remain of anthropological interest (see Chapter 7; Phull *et al.* 2015). In the last two decades, scholars have employed conviviality in a much broader sense, deriving from the Spanish words *convivir* (to live together/to share the same life) and *convivencia* (a joint/shared life) (ibid; see also Chapter 11) yet to a degree that it becomes tantamount to sociality.[1]

Commensality, or eating together, is perhaps one of the most salient instances of human sociality (Fischler 2011: 529). Its literal meaning is eating at the same table (*mensa*), but commensality is often more broadly defined as 'eating with others' (Sobal and Nelson 2003), 'eating in groups' (Fischler 2011), or 'eating and drinking together in a common physical or social setting' (Kerner *et al.* 2015). The Lord's Supper or ritual feasts, such as the Indonesian *slametan* (Geertz 1960), are classic examples of commensality. Yet the scholarly interest in commensality does not derive solely from its practice in religious, sacred, or ritualised contexts but also recognises it in its quotidian form of the shared common meal because of its potential to create, maintain, and renew social ties and thus group belonging. Indeed, Georg Simmel postulated as early as 1910 in 'The Sociology of the Meal' (*Die Soziologie der Mahlzeit*) that 'the immense sociological significance of the meal' lies in the fact that '[p]ersons who share no particular interests can find themselves sharing a meal'(Simmel 1994 [1910] in Symons 1994: 346). For sociological and anthropological analyses, it is of interest what such practices do (to commensals and non-participants, to their surroundings) and in what ways, through their comparable particularities (who eats with whom, where, when, how, and what). As a result, it would be a fallacy to view commensality solely as fostering social cohesion since it always also includes hierarchical dynamics among commensals and social processes of exclusion for those not taking part (Bloch 2005; Fischler 2011; Pitt-Rivers 1977). Notably, commensality is – like sociality and conviviality – a social process, and we hold that the senses play a key role in understanding this 'dynamic relation matrix'.

For instance, Martin Manalansan (2006) highlights in his study of Asian American communities in New York how the immigrant body is 'culturally constructed to be the natural carrier and source of undesirable sensory experiences and is popularly perceived to be the site of polluting and negative olfactory signs' (2006: 41). While New York City is 'almost always visually represented by the mythical image of an odorless Manhattan skyline', smells proliferate, and the city is 'an arena for contesting, creating and imposing regimes of olfactory meanings and corporeal practices' (2006: 43). Here, the trope of the 'smelly immigrant' leads Asian Americans to feel obliged to contain their sensual presence through domestic food preparation and consumption to lessen the impact of cultural and economic difference (Manalansan 2006: 47). Yet these 'struggles of containment and domestication of food aromas around Asian immigrant homes are part of competing economic and political interest in global capital in perpetuating racial and class subordination (Sassen 1996)' (ibid).

As with this example, in this book we are particularly interested in those dynamic social processes in which people engage with food in urban spaces and contend that food and the senses play a crucial role in how people interact with each other; create sociality and conviviality; and, in this process, also shape the city and are shaped by it. Such processes are, of course, also temporal and are fashioned as much by people's imaginary futures as by their particular individual and collective pasts.

Food and memory

Food's sensuous capacities make it ideal for the recalling of memories – often involuntarily (Holtzman 2006; Seremetakis 1996; Sutton 2001). Central within the burgeoning memory scholarship is thinking about the relationship between the past and the present, in which previously history and memory are juxtaposed as objective versus subjective pasts (Nora 1989; Ingold 1996). Yet neither history nor memory is reliable for accessing 'the true past' because pasts are 'always presentations, always constructions' (Hodgkin and Radstone 2003: 2). Since history and memory are mutually informed in their remaking, '[t]o ethnographically explore the fluid, interdependent relation between history and memory' renders history contextual, 'and "memory" whether it is collective or individual, becomes a dimension of intersubjective significance' (Birth 2006: 177). Importantly, because memories are created through experiences in the past, they also shape our experiences in the present, and thus, memory becomes 'intrinsically linked to identity' (Antze and Lambek 1996: xii; Assmann 2006: 7). Food and memory are intricately connected in creating and maintaining ethnic identity, although memory here is often implicit rather than used as an analytical tool in food studies (Holtzman 2006: 366).

Nevertheless, anthropology's fascination with memory is perhaps due to its project of 'understanding continuity' (Berliner 2005: 205) or how exactly social cohesion is achieved throughout time in socio-cultural transmission across generations (Bloch 2005) – despite or because of social change. Regarding the experience of food, memory becomes embodied (Stoller 1997), which renders food practices particularly powerful social transmitters – akin to Bourdieu's (1977) *habitus* or Connerton's (1989) body memory (see also Holtzman 2006). C. Nadia Seremetakis (1993), for example, demonstrates how food and the senses play crucial roles in the relationship between grandmother and child in Greece. She likens raising the child to baking bread, where 'Baking gives form: color, shape, texture. Enculturation is a sensory process and tied to the acquiring of form. It draws its imagery (color, shape, texture) from the body and food processing' (1993: 3). She describes how '[b]abies are wrapped in cloth, and the dough is covered with blankets and towels to rise. The mouth of the grandma (softening the bread) is an oven, as is the womb (see also duBois 1988)' (ibid).

She then also charts the passage from city to country that the child takes to visit the grandmother, intricately linked to food and the senses:

> Every station was identified with specific foods, their particular tastes and smells – one station with *souvláki*, another with rice pudding, pistachios, *pastéli*, dried figs. The child traveled through substances to reach grandma; a journey that sharpened the senses and prepared the child for diving into the village. The child arrived to the smell of the ocean, the trees, lemon, orange, olive, the sound of the donkey's bray, and the omnipresent, loud, loud music of the cicadas: sensory gates that signified entry into a separate space.
>
> (1993: 5)

This stunning prose alerts the reader to the flows of the senses and how they become memory – 'the migration of sensory forms via material artifacts, and the memory they leave behind' (1993: 7), vis-à-vis Greek modernity and its concomitant losses.

Perhaps unsurprisingly, much recent focus in the anthropology of memory has been directed towards better understanding nostalgia (Angé and Berliner 2014; Todorova and Gille 2010). Nostalgia is often perceived as looking back at the past through rose-tinted glasses, a trivialising sentimentality that idealises a past. Originally coined in 1688 by the Swiss medical student Johannes Hofer, the term *nostalgia* combines the Greek words for 'return to the native land' (*nostos*) and 'suffering or grief' (*algia*) (see Boyer 2006; Hirsch and Spitzer 2003). Hofer's intention was to distinguish a fatal or near-fatal physiological disorder, suffered by displaced soldiers at the time, from the notion of homesickness. Patients were believed to be cured by returning them to their origins. Today, nostalgia is no longer used to describe a disease in need of curing, but – just like its bigger sibling, memory – is applied in various ways. These different conceptions of nostalgia have as common denominator their 'link with absence or removal from home', which has also been broadened to a general sense of loss (Hirsch and Spitzer 2003: 82). Gastronomic or culinary nostalgia (Swislocki 2009) is thus often associated with diasporic or expatriate communities and their experience of displacement (Holtzman 2006).

The chapters in Part II of this book deal explicitly or implicitly with nostalgia or absence/removal from home as 'personal consequences of historicising sensory experience which is conceived as a painful bodily and emotional journey' (Seremetakis 1996: 4). Many illustrate how such a sensed (temporary or permanent) loss of home manifests in food and foodways that simultaneously serve as strategies to recuperate an experienced or imagined past and, in this process, create new socialities, convivialities, and commensalities to 'feel at home' in the city.

The city

The senses, while intangible, have tangible effects; the sensory environment produces strategic responses to moving around the city. Or, put differently, the senses are skills that we employ to interpret and evaluate the world (Ingold in Vannini *et al.* 2012: 15). Cities are places of heightened sensory experiences where urban dwellers evaluate and create their everyday lives through sensoria (Degen 2008; Rhys-Taylor 2017). People's engagement with food both shapes cities and is shaped by them, where acknowledging the senses in urban food practices is essential to link people both to each other and to a place. Daily encounters may seem insignificant yet pack a persuasive punch; examples include people avoiding certain places deemed noisy or smelly, seeking certain sensory memories by eating or shopping at certain restaurants or food stores, or hearing racist apologetic complaints about an apartment still smelling of Indian food due to its previous tenants. Such microscopic experiences are not only the sensibilities of how people connect and produce city life; they are also part of the socio-political fabric of the city, where the proliferation of new social networks based on lifestyle and other interests is changing how we live in cities.

Increasing proximity also heightens sensory incursions where cities bring us 'together apart' and living in close proximity to strangers produces a register of experience that has tangible, tactile consequences: the sound, smell, and sight of others are part of the white noise of life that can erupt into neighbourhood disputes. Triggered by sensory input, spatial conflicts are part of everyday urban life, and their ruptures are widely felt. Hostility to outsiders and people who appear different and eruptions of feeling over diverse urban sensory stimuli contribute to produce the micro inclusions and exclusions of everyday urban tactics. For Lauren Berlant (2008), city spaces are spheres of intimacy in which feelings towards the self, others, and the nation can be experienced in formative ways. While creative city rhetoric promotes diversity, its acceptance is highly contingent and subjective.

While the everyday practices of city dwellers have become accepted as crucial to understand in anthropology and other disciplines, the role of the senses has been less explored. Two approaches dominate the analysis of city life: one is oriented to urban policy and design and proclaims a deliberate 'top-down' shaping of social structures (for example, Weber 1958 [1921]), and the other asks how the city is experienced, absorbed, interpreted, and evaluated by individuals and groups from the 'bottom up', such as work initiated by Simmel (1976 [1903]), De Certeau (1984), and Ronan Paddison and Eugene McCann (2014). Within planning, a search for spatial order and disciplinary control has dominated throughout history, and the senses are often acknowledged as a nuisance: for example, smell and noise from urban livestock, such as cattle lots or chicken farms. Indeed, such complaints drove the instigation of urban regulation, prompting the 'removal' of nature from the city (see Brinkley and Vitiello 2014).

Paying attention to the everyday city was initiated by Michel de Certeau and Henri Lefevbre and served as a crucial starting point for the understanding of cities as places where everyday experiences and sensorial regimes shape socio-political subjectivities and processes. Indeed, Lefebvre (1996) suggested we can only fully comprehend a city when we consider how the material, imaginative, and experiential dimensions of urban life intersect and play out in the lives of its inhabitants. Just as Lefebvre (1991) argued that space is permeated with and produced by social relations, we argue that these social relations are as much informed by the sensorium and sensory experiences. Several authors have shown that space is more than merely geographical; it creates and is created by sensuous regimes and practices (see Degen 2008; Law 2001; Manalansan 2006; Rhys-Taylor 2017).

However, we do not want to juxtapose the city seen from above and from below. Instead, we seek to hit a middle zone where we recognise that cities are articulated through many parts, including embodied, social, cultural, demographic, climatic, and historical elements. We need to understand the city from the experience of its dwellers (Rhys-Taylor 2017). For geographers, Yi-Fu Tuan (1977: 18) argues that 'an object or place achieves concrete reality when our experience of it is total, that is, through all the senses as well as with action and reflective mind'. Ingold has proposed to understand place by seeing the environment not as the surroundings of an organism but as a zone of entanglement (2008: 1797). He suggests life is made of movements, of strands that become tied up with each other, something he calls a 'meshwork'. Such a proposition is useful to think about the ways places are much more about the fluidity of who and what moves and connects them than about the space itself as fixed.

Place-making and sensing food in the city

Place-making, in almost all chapters in this book, challenges the idea of place, identity or belonging as being fixed and stable. Places are not only imagined by other places and times; they are also made from interactions with the outside (Massey 1992). While the idea of a closed space is untenable, according to Doreen Massey, we do observe the ideas of localism and authenticity as recurring themes in many current foodways (Beriss 2019). Such foodways create a certain notion of a locality with a stable and internalised history and, at the same time, are often highly changeable responses to 'McDonaldisation' (Ritzer 1993) – the globalisation of food (Beriss 2019: 64). An example is the work of Melissa Caldwell (2002, 2004), who writes about Muscovite identity construction and Russian food. Encountering an increasingly transnational commodity market, locals have linked their personal food experiences with broader political issues, re-appropriating foreign cuisine in a way that reflects a socialist ethics of sociality and collective responsibility (see Caldwell 2002).

The French concept of *terroir*, which attends to 'the geology and climate of particular regions and to the putative way in which they combine to help produce agricultural products' (Beriss 2019: 63), is perhaps the most prominent example of linking locality and food production. Amy Trubek (2009) explores this concept further since 'the origins of food and beverage determine prevailing notions of taste' (2009: 8). While a shift to 'the local' has become a catch cry of the sustainable food movement (see literature on alternative food networks, including Goodman *et al.* 2012), the role of taste within that shift is not so well explored. Advocates for Slow Food – and its related branch of slow cities – are perhaps the strongest proponents to draw the senses into this conversation, where they honour and uphold food values of pleasure, conviviality, and quality (see Petrini 2004; Pink 2008; Siniscalchi 2014). Trubek investigates 'terroir' as 'the notion that the natural environment can shape the taste of wine', in which further refinement as 'knowing' and 'discerning' recognises how cultural context also tweaks this definition, yet terroir is not easily applicable to produce in other cultural contexts, such as Darjeeling tea (Besky 2014) or Palestinian olive oil (Meneley 2011) (see Beriss 2019: 63). Notably, Trubek acknowledges how the 'commodities are not perceived as sensual objects, capable of evoking pleasurable and meaningful moments' (2009: 15), suggesting that a (re)connection to place can revitalise an ability to taste food in its fullest capacity.

The role of the senses in food and place-making and the processes of urbanisation has been excellently described by Lisa Law (2001). Law links the senses to space, place, and culture in her study of Filipino domestic workers in Hong Kong, where they (re)create a 'sense' of home. She describes how domestic workers return one day each week to Little Manila, where they recover from sensory reculturation after working in Chinese homes to revisit their Filipino culture, food, and friends whilst finding new ways to engage with the city on their own terms. The convivial consumption of Filipino food in Little Manila through 'its taste, texture and aroma' helps embody and emplace these women as new national subjects. Law's research reminds us of the political implications of food, senses, and place, where food practices act as a cultural mediator to ascribe identities to marginalised participants who come together in the city as a site of resistance.

Another relevant example is Manpreet Janeja (2010), who ascribes affective agency to food in her analysis of Bengali cuisine across borders. She explores the 'foodscape', defined as produce and food practices that create a sense of place or '*desh*'. In Sutton's review of her work, he identifies one illustrative phrase where a dish of mashed chillies '*attaches* itself to the mistress . . . touches her, wraps her in its embrace, dissolves her bodily boundaries, and enters her. It *makes*: her eyes water, her face turn red, her cough, and her lose her temper' (Janeja 91; emphasis in original; in Sutton 2013).

Jean Duruz's work on Singapore (2006, 2011) provides an important touchstone to consider 'food on the move' and how it relates to identity, place, culture, and memory. Her 2006 article literally shifts in materiality and identity construction between Australia and Singapore to explore concepts of home with the senses of associated dishes. Drawing on accounts of a couple from different cultural backgrounds and experiences, the article examines how 'place-making processes intersect' to de-stabilise traditional nationalistic identities while describing new ways of homemaking in which 'one can yearn for the comforts of the home-in-memory and those of home-in-imagination' (Duruz 2006: 105). Duruz's closing argument is that we should not 'simply celebrate this "different" table; instead, we should examine its ambivalent economies of comfort and discomfort, as well as the effort, imagination and will needed to make it work' (2006: 114). This 'food on move' perspective is further embellished in her 2011 article, in which she examines laksa's arrival in Adelaide and its later normalisation as 'one of Australia's "borrowed" foodways' (2011: 605) to explore questions of belonging.

Anna Mann's (2015) thesis examines mundane spaces of consumption in a range of western Europe settings, including sensory science laboratories, restaurants, hospitals, at a wine tasting event, daily dinners and as a meal in a convent. She argues that 'tasting is a *physiological response* to a food object, leading on to a *multi-sensory experience* of its qualities, that do not just emerge from the food but are co-shaped by the *context* and that give rise to sensorial *knowledge*' (emphasis in original).

As Alex Rhys-Taylor (2017) elaborates in his beautiful study *Food and Multiculture: A Sensory Ethnography of East London*, the way the senses work

> happens at a relatively microscopic, interpersonal level, invisible at the scale of the whole city. So microscopic, in fact, that such experiences are often felt to be of barely any sociological significance for the individual, let alone the broader mass of cities or societies.
>
> (2017: 2)

Like him, we contend that the senses shape the city and also have significant consequences for the urban socialities of which they are part.

As this literature shows, urban environments are complicated, layered, and messy, and they intersperse with the world around them. They are sensed in various, multiple ways; they include and exclude; they are shaped by and they shape sociality. In this book, we do not aim to define cities as demarcated spaces that bring along specific kinds of senses or foodways. Cities neither have clear physical boundaries, nor do their inhabitants fix their lives in them. Where does the city end and urban sprawl begin? How to grasp seasonal migration? What about all those places that are neither one nor the other? Where one detests the noise of bars, the other sees it as an expression of a lively neighborhood. What one person relates to a certain sight, sound, or smell can be completely different for someone else or at a different time. While not resorting to a completely individual experience, we acknowledge the ways in which everyday experiences and senses shape place and how place shapes experience and senses in return. The senses create our sense of place as it is 'known, imagined, yearned for, held, remembered, voiced, lived, contested and struggled over' (Feld and Basso 1996: 11).

We do not aim to define the city, let alone to cover the urban, in a systematic way. By looking at food and foodways, we aim to see the city as a place of imagination and everyday practices. 'Cities are everywhere and in anything', above, below, and between the surface, Arjun Amin and Nigel Thrift contend (2002: 7). We propose that in order to understand the social questions that connect emerging urban food practices, we need to take account of the everyday experiences and sensibilities that inform them. A focus on the senses provides an understanding of the ways in which the physical and social environment is being reshaped in relation to food, memory, and materiality and how these aspects connect past, present, and future in local epistemologies (Seremetakis 1996).

An overview of the book sections

While many of the chapters could fall within multiple sections, the categorisations chosen here aim to draw out, refine, and deepen specific aspects and understandings of 'being at home', memory, belonging, boundary-making, gender, and class, grounded in ethnographic research. The book is divided into three parts:

Part I: The city and its other

Organic food shops, foodie tours, urban gardening, farmers' markets, food activism: cities around the world have seen considerable transformations in terms of foodways during the last decades. Such changes could partly be pinned down to a global, mostly urban, renewed attention to food, health, and environment, which is now gaining more momentum in many countries, reinforced by the recent COVID-19 pandemic. To understand the subjectivities of such urban phenomena, we need to recognise the experiences of city dwellers (Rhys-Taylor 2017) and how they embody the various practices related to change. We recognise how cities, their inhabitants, and the way they are experienced and imagined are constantly in flux. Activism, the attention to gardening and beekeeping and exploring a different side of the city seem to be moments of exploration of another possible city – a city that is better, healthier, fairer, or sensorially more pleasant. These ideoscapes of something better often seem to revert to something opposite to the city: the countryside – or nature.

Cities exist in relation to their antipode, the countryside, nature, or nonhumans: cities rely on food production outside the city, migrants moving in and out, urban imaginations of what the countryside is like. Urban dwellers create their ways of relating to and imagining from the countryside. The idea of the countryside as the place with the purest and most authentic foodstuff has deep roots (Domingos *et al.* 2014: 2). Certain food practices of working on the land and being in touch with the soil seem to emerge out of a feeling of disconnectedness from the food chain. The practices seem at first hand to be a counter-narrative of the modern city, of taking a stance against the problematics and unsustainability of modern city life and inequality and seeking a way to relieve urban ills (Imbert 2015). However, these urban practices can only take shape in relation to city life, rearticulating country life worlds in the lifestyle politics and cultural identities of urban dwellers. This tension of the city and the countryside also shows itself in the increased blurring of these spaces: migration between the two – cities becoming greener and countryside more and more incorporated in the urban sprawl (Imbert 2015) – or, more relevant for this book, in the foodways of city consumption. In this section of the book, we ask how the senses create a sense of being connected to its other – the soil, the 'true' city, or the countryside – and therefore to the city? What do these imaginations and practices of working on the land and with the soil do for making sense of the city? How do food practices create imaginations of the rural; of the city; of the past, present, and future? How is this image of the other and of the city created through sensorial experience?

In Chapter 2, Vincent Walstra discusses an urban gardening project in which the return to the soil, the touching with the hands gives a new turn to city life. Realising how out of touch with nature one is, the people working on the urban farm experience a hands-on way that turns against the industrial food system and instead appreciates sensorial experiences. Walstra argues that this reveals a societal desire to reconnect with nature and food through the senses. Somewhat similarly, in Carole Counihan's work (Chapter 3), sensory bodily engagement enables consumers to connect with farmers, land, and food production. Counihan describes Cagliari food activists and ways of somatic actions in working on the land and countryside. She shows how this corporeal involvement with food challenges disjunctions of the contemporary city. The corporeal practices bring together self and body, consumption and production, and city and country and therefore support the goals of food activism for a more just, sustainable, and healthy food system. In Chapter 4, Ferne Edwards brings in another important 'other' of the city: that is, nonhumans, or animals in this case. Chapter 4 explores sense-making between humans and bees and pushes the idea of displacing human-centricity in ethnographic knowledge production. The Australian beekeepers in Edwards's chapter need the skill to sense their bees, but bees also sense them, and this mutual sensing creates human and nonhuman relationships in cities.

Chapter 5 has a different methodological approach to attend to the sensorial explorations of the city within. Roos Gerritsen shows through a combination of a photo-essay and a written text how a group of local food lovers in the city of Chennai explore neighbourhoods to experience the real city. Their busy urban lifestyle and lack of time and possibility to *be* in the city make them interested in exploring neighbourhoods together. Gerritsen argues that through these explorations, the foodies explore what they say is the genuine, real, or vernacular city. The photographs tell their own story and aim to depict the fragmented food tours as well as the ways in which the sensorial experience creates the spaces in which they take place.

Part II: Past in the present: memory and food

In 1991, during the great transformations that swept across the former Eastern Bloc after the fall of state socialism, Slavenka Drakulić (1993) recounts her disappointment that the brighter future for which she had hoped did not materialise. Instead, after Slovenia's secession from Yugoslavia, war ensued. She feels cheated but suddenly catches herself thinking about peaches:

> We thought that after the revolution peaches would be different – bigger, sweeter, more golden. But as I stood in line at a stall in the street market I noticed that the peaches were just as green, small and bullet-hard, somehow pre-revolutionary.
>
> (1993: xii)

Perhaps around the same time, Nadia Seremetakis (1996) longs for *rodhankino*, the particular peach of her childhood that is different to any peach she can find in the USA. In her summer trips to Greece, she notices that this variety of peach has disappeared. Disappointed about her search for this peach in vein, she realises that the peach has become narrative – a living memory made social through its recounting in remarks about its flavour: 'nothing tastes as good as the past' (1996: 1).

Both women's stories centre around a peach – its actual materiality, but also the peach as sensory imagination and for what it stands in. Drakulić's peach becomes representative of the failure of the socio-economic improvement she had imagined would arrive with the political transformation – an unattained ideal future home. Seremetakis's peach represents her childhood, associated with particular tastes of home – a past that is as much gone as the peach that has vanished. The two different peaches illustrate nicely food's inseparable entanglement with the senses, memory, and history (Seremetakis 1996; Sutton 2001; Holtzman 2006) that enables it to travel across time and space, and Part II engages specifically with peoples' notion of what home means, their longing for it, and varying attempts at (re)creating home or making a new home. In Chapter 6, Joel R. Hart and Daniel Monterescu follow *amba*, a mango pickle condiment, on its journey across time and space. While for Iraqi Jews in London, *amba* exemplifies culinary nostalgia – a longing for an unattainable home – meanwhile, *amba* has become a popular condiment for both Israeli and Palestinian street food. Yet its particular distinct garlicky odour lingers with its (commonly male) consumers and renders it contentious. Food, as a crucial tool for creating senses of belonging and home, can enforce differences but also challenge and subdue notions of gender, class, ethnicity, kinship, and nation and, in turn (re)shape cities. Since a sense of belonging is a processual attainment that depends in part on familiarity, this familiarity can be lost and longed for either through leaving home (see Chapters 6, 8, 9 and 10) or through rapid and dramatic transformation processes in one's home, as is the case in Chapter 7. Grit Wesser here explores how Thuringians attempt to recuperate a simpler but emotionally laden sense-scape of their socialist past through the consumption of Thuringian festive cakes. She shows how discourses and perceptions of women's domestic labour of love in producing homemade cakes are extrapolated to the regional home or *Heimat*. 'Homemade' cakes' aesthetic, gustatory, and tactile qualities are subtly altered to achieve convivial commensality at life-cycle celebrations that foster a sense of belonging beyond the familial home without explicit references to German nationhood. Crucially, as most chapters show, how home was sensed becomes fixed in our imagination and a return foreclosed because time and space are constantly changing. This is particularly pertinent in cases of migration since migrants often carry cultural food practices on their journeys to create new senses of 'being at home'. Shuhua Chen (Chapter 8) shows through a tour of a marketplace with her rural migrant friend in Shantou, Southern China, that rural migrants can experience home there briefly through a sensory totality or synesthesia (Sutton 2010). The market place also reflects how industrialisation and urbanisation with their concomitant migration have altered it and, in turn, shaped the foodways of its dwellers – locals and migrants. Diti Bhattacharya (Chapter 9) highlights how two Bengalis living together in Australia negotiate the differences and commonalities in Bengali cuisine that stem from their varying culinary memories and practices as a Muslim Bengali from Dhaka and a Hindu Bengali from Kolkata. The creation of a new home in a city away from home thus draws not just on one home but on the culinary experience of two cities, which are explored, combined, and re-imagined. Premila van Ommen (Chapter 10) explores in her digital ethnography how young British Nepali celebrate culinary traditions online, in which 'being at home' can refer to visits to Nepali restaurants in London but is often also expanded through the inclusion of other Asian culinary traditions. Through such digital practices, they make their taste public (Counihan and Højlund 2018) and create an online network of Nepali diaspora that constantly shapes what it means to be a British Nepali today through snapshots – photographs and video stories – of their culinary practices.

Part III: Disrupting and re-imagining

Perceptions of our cities and diets are constantly in flux. Increasing urbanisation is often perceived as a negative and unstoppable force as cities throughout the world grow in size, density, and noise, producing side effects of pollution, high resource consumption, commercialisation, and crowding. However, the city has always changed its form, purpose, and perspective throughout history. Jacob Klein (2014: 4) reminds us of the relationality between food and cities, asking:

> [H]ow should we approach the relationship between changes and continuities in food consumption habits, on the one hand, and changes in social boundaries, roles and relations, on the other? And in what ways is our understanding of the social dynamics of food consumption in specific

localities, especially in recent decades . . . illuminated by comparative perspectives and an attention to wider, even 'global' processes?

Recognising the growing power of cities, this section explores the push and pull of cities to where they go beyond their status quo to reach 'new normals'. This section asks how people are re-imagining their cities – looking forward, rather than back – through food and the senses to explore new boundaries, roles, and relationships that can be fostered towards urban food futures. It asks, quoting Melissa S. Biggs (this volume), 'What does change taste like? What are its smells, its sounds?'

The 'push and pull' of urban forces can be both detrimental and advantageous. While for some, change can be confronting; for others, the process of disruption symbolises the goal of changing the status quo. Recognising persistent inequalities or injustices while often seeking to instil new social justice and environmental values, food practices are being used as a tool to engage, protest, and imagine preferred ways of being together through food. These changes may be externally forced or internally driven, with two sharp contrasts in the re-imagining of the city present in this section: the first where factors of capitalism, profit, and planning reign (Chapters 11, 12 and 13) and the second where the citizens assert their presence through the senses to reclaim their city and the future direction in which they would like to take it (Chapter 15). However, more subtle understandings of re-imaging the city are also occurring, such as by reconnecting city and country through food and the senses (Chapter 16).

In Chapter 11, Melissa S. Biggs associates the gastronomical concept of 'palate' to particular cities. Her research notes the sensual nuances across place as emerging tastes are shared and stick fast, focusing on 'tapatio' tastes, the term given to natives of Guadalajara. Biggs's chapter forms the beginnings of a 'cultural chronology' (Paterson *et al.* 2016) of the senses, a theme shared by Monica Stroe (Chapter 12) and Aitzpea Leizaola (Chapter 13) to Bucharest and Basque Country respectively. Stroe's research focuses on the celebration of Labour Day at the Obor Market in Romania as a public arena for the appropriation and articulation of taste, where the privileged mobilise their cultural capital to re-purpose working-class cuisines. Her research focuses on the altered consumption of *mici*, traditionally an affordable, working-class street food consisting of freshly grilled, minced meat rolls. In Romania, street food has joined global gastronomic trends in keeping with Sharon Zukin's (2008) 'latte towns', witnessing the spatialisation of rising inequality as gentrification. In her chapter, Stroe argues that middle-class consumers take on identities of 'foodie *flâneurs*', and 'focused on the aesthetics of the city, engaging with street food's sensory cues, they experience the city with a sense of detachment, as spectators to its many facets, seeking to acquire and perform culinary capital' (Stroe, this volume). Leizaola (Chapter 13) examines this process of change based on the consumption of the *pintxo* in Donostia, the capital city of Basque Country, a region known for its gastronomic heritage. In contrast to the foods discussed by Stroe, this cuisine has a background in class and wealth distinctions; the *pintxo* represents a one-bite delicacy of high culinary standing that allows one 'to taste the latest creations in culinary innovation for a reasonable price' (Leizaola, this volume). Here, the discourse twists to 'local versus tourist or foreign' consumption, and traditions have changed to accommodate the unaware cultural palate. Alternatively, Catherine Earl (Chapter 14) examines urban-rural change and reconnections. Her research describes the experiences of a Ho Chi Minh City family – the parents are first-generation migrants; the children are city born. She asserts that 'A dish does not belong to a place. Rather, it belongs to the senses; it is situated in its practice' (this volume) where 'cultural capital is mobile and can be sourced and accrued in diverse places' (ibid). The final chapter explores how people are actively shaping the city's imaginary through experiencing senses in food preparation and consumption. Paz Saavedra *et al.* depart from a focus on the pleasurable senses to acknowledge discomfort and effort in the creation of a traditional dish to sell at an agroecological market in Ecuador. Their chapter develops a theme of solidarity and an ethics beyond the human where people work together to uphold traditional customs whilst supporting environmental goals in the face of agro-industrial pressures to carve out new spaces in the city where their goals can be shared with others.

While not all practices described in the chapters directly engage with the city, their practices influence how city people live while acknowledging the material flow from rural to urban. Indeed, we recognise that no city is an island. Instead, cities can provide a common space where people can come together to share knowledge; to stand their ground; to challenge the status quo; to respond to crisis; and, at times, to change laws and policies. 'Re-imagining', then, considers how the senses can help us redesign our urban food futures. What futures can we imagine? Who is considered in this redesign? How can we use our senses to feel what this change may mean? This final section argues that sensing food can be both a tool for change and solidarity to make sense of future cities.

Conclusion

In this introductory chapter, with its brief review of relevant anthropological works, we have positioned ourselves within sensory anthropology and how it can be fruitful to the study of food and foodways in urban contexts. We do not merely follow Howes in his cultural approach, nor do we take a purely phenomenological approach (see debate in *Social Anthropology* between Pink and Howes 2010; Ingold 2011). Instead, we see merit in both because individual experience is made by its surrounding, and this surrounding is made by socio-cultural contexts. As a result, the ways in which the senses create and are created by experience and context requires anthropological attention. We follow Vannini *et al.* (2012) in acknowledging the engagement and fuzziness of sensorial experiences. Sensorial experiences are particularly heightened in food practices because food is more than a nourishing materiality: invested with sensual qualities, it is rendered an emotional force and an embodied memory that enables the transmission of cultural values across time and space. Therefore, we see food as one of the most significant social conductors that people engage daily in – out of necessity but also out of pleasure – in either producing, processing, consuming, or disposing of food. Situating food practices within the city further intensifies relationships between people, food, and the senses due to physical proximity and cultural diversity. The senses are one way in which to navigate and unpick such saturated terrains in which we see the city as diverse, dynamic, fluid, and entwined, where people create the city and the city influences people's lives. Food and the senses have the potential to bring people together and to connect them to place, whilst retaining a lingering power that helps refine and define identities, connect them to past and present, and stay with people 'at home' or move with them as they adjust to new environments. The chapters in this volume address this nexus of 'food, city, and the senses' to demonstrate how inquiry into food and the senses can help us more fully understand the vibrant life of the contemporary city.

Note

1 Paul Gilroy's *After Empire* (2004) heralded this broader use of conviviality, which is increasingly popular in studies of migration and intercultural relations for understanding superdiverse urban context (for a brief overview, see Wise and Noble 2016).

References

Alexander C., Gregson N. and Gille Z. (2013) 'Food Waste', in Murcott A., Belasco W. and Jackson P. (eds) *The Handbook of Food Research*. London and New York: Bloomsbury: 471–484.

Amin A. and Thrift N. (2002) *Cities: Reimagining the Urban*. Cambridge: Polity Press.

Angé O. and Berliner D. (eds) (2014) *Anthropology and Nostalgia*. Oxford and New York: Berghahn Books.

Antze P. and Lambek M. (1996) *Tense Past: Cultural Essays in Trauma and Memory*. New York and London: Routledge.

Assmann J. (2006) *Religion and Cultural Memory*. Stanford, CA: Stanford University Press.

Berlant L. (2008) *The Female Complaint: The Unfinished Business of Sentimentality in American Culture*. Durham, NC: Duke University Press.

Berliner D. (2005) 'The Abuses of Memory: Reflections on the Memory Boom in Anthropology', *Anthropological Quarterly* 78 (1): 197–211.

Beriss D. (2019) 'Food: Location, Location, Location', *Annual Review of Anthropology* 48: 61–75.

Besky S. (2014) *The Darjeeling Distinction: Labor and Justice on Fair-Trade Tea Plantations in India*. Berkeley: University of California Press.

Birth K. (2006) 'The Immanent Past: Culture and Psyche at the Juncture of Memory and History', *Ethos* 34 (2): 169–191.

Bloch M. (2005) 'Commensality and Poisoning', in *Essays on Cultural Transmission*. Oxford: Berg: 45–59.

Bourdieu P. (1977) *Outline of a Theory of Practice*. Cambridge, UK: Cambridge University Press.

Bourdieu P. (1984) *Distinction: A Social Critique of the Judgment of Taste*. Translated by Richard Nice. Cambridge, MA: Harvard University Press.

Boyer D. (2006) 'Ostalgie and the Politics of the Future in Eastern Germany', *Public Culture* 18 (2): 361–381.

Brinkley C. and Vitiello D. (2014) 'From Farm to Nuisance: Animal Agriculture and the Rise of Planning Regulation', *Journal of Planning History/Society for American City and Regional Planning History* 13 (2): 113–135.

Bull C. J. C. 2018 (1997) 'Sense, Meaning and Perception in Three Dance Cultures', in Howes D. (ed) *Senses and Sensation: Critical and Primary Sources, Vol. 4: Art and Design*. London: Bloomsbury: 263–276.

Caldwell M. (2002) 'The Taste of Nationalism: Food Politics in Postsocialist Moscow, *Ethnos* 67 (3): 295–319.

Caldwell M. (2004) 'Domesticating the French Fry McDonald's and Consumerism in Moscow', *Journal of Consumer Culture* 4 (1): 5–26.

Castaing-Taylor L. and Paravel V. (2012) *Leviathan*. Film. USA.

Classen C. (1997) 'Foundation for an Anthropology of the Senses', *International Social Science Journal* 49 (153): 401–412.

Clifford J. and Marcus G. E. (eds) (1986) *Writing Culture: The Poetics and Politics of Ethnography*. Berkeley: University of California Press.

Connerton P. (1989) *How Societies Remember*. Cambridge, UK: Cambridge University Press.

Counihan C. and Højlund S. (eds) (2018) *Making Taste Public: Ethnographies of Food and the Senses*. London: Bloomsbury.

Cox R., Irving A. and Wright C. (eds) (2016) *Beyond Text? Critical Practices and Sensory Anthropology*. Manchester: Manchester University Press.

De Certeau M. (1984) *The Practice of Everyday Life*. Berkeley: University of California Press.

Degen M. (2008) *Sensing Cities: Regenerating Public Life in Barcelona and Manchester*. London: Routledge.

Domingos N., Sobral J. M. and West H. (eds) (2014) *Food Between the Country and the City: Ethnographies of a Changing Global Foodscape*. London and New York: Bloomsbury.

Douglas M. (1982) *In the Active Voice*. London: Routledge.

Drakulić S. (1993) *How We Survived Communism and Even Laughed*. New York and London: Harper Perennial.

duBois P. (1988) *Sowing the Body: Psychoanalysis and Ancient Representations of Women*. Chicago: University of Chicago Press.

Duruz J. (2006) 'Living in Singapore, Travelling to Hong Kong, Remembering Australia: Intersections of Food and Place', *Journal of Australian Studies* 30 (87): 101–115.

Duruz J. (2011) 'Tastes of Hybrid Belonging: Following the Laksa Trail in Katong, Singapore', *Continuum* 25 (5): 605–618.

Evans D., Campbell H. and Murcott A. (eds) (2013) *Waste Matters: New Perspectives on Food and Society*. Hoboken, NJ: Wiley-Blackwell.

Evans-Pritchard E. E. (1940) *The Nuer*. Oxford: Clarendon Press.

Featherstone M. (2010) 'Body, Image and Affect in Consumer Culture', *Body & Society* 16 (1): 193–221.

Feld S. (1991) 'Sound as a Symbolic System: The Kaluli Drum', in Howes D. (ed) *The Varieties of Sensory Experience: A Sourcebook in the Anthropology of the Senses*. Toronto: University of Toronto Press: 79–99.

Feld S. and Basso K. H. (eds) (1996) *Senses of Place*. Santa Fe, New Mexico: School of American Research Press.

Feld S. and Brenneis D. (2004) 'Doing Anthropology in Sound', *American Ethnologist* 31 (4): 461–474.

Fischler C. (2011) 'Commensality, Society and Culture', *Social Science Information* 50 (3–4): 528–548.

Geertz C. (1960) *The Religion of Java*. Chicago: University of Chicago Press.

Geertz C. (1973) *The Interpretation of Cultures*. Boston: Beacon Press.

Geurts K. (2003) *Culture and the Senses Bodily Ways of Knowing in an African Community*. Berkeley: University of California Press.

Ghertner D. A. (2011) *Rule by Aesthetics: World-Class City Making in Delhi*. Oxford: Oxford University Press.

Gibson J. J. (1979) *The Ecological Approach to Visual Perception*. Boston: Houghton Mifflin.

Gilroy P. (2004) *After Empire: Melancholia or Convivial Culture*. London: Routledge.

Goodman D., DuPuis M. and Goodman M. (2012) *Alternative Food Networks: Knowledge, Practice, and Politics*. Abingdon: Routledge.

Goody J. (1982) *Cooking, Cuisine and Class: A Study in Comparative Sociology*. Cambridge: Cambridge University Press.

Grimshaw A. (2007) 'Reconfiguring the Ground: Art and the Visualization of Anthropology', in Westermann M. (ed) *Anthropologies of Art*. Williamstown, MA: Sterling and Francine Clark Art Institute: 195–220.

Hirsch M. and Spitzer L. (2003) '"We Would Have Not Come Without You": Generations of Nostalgia', in Hodgkin K. and Radstone S. (eds) *Contested Pasts: The Politics of Memory*. London: Routledge: 79–95.

Hodgkin K. and Radstone S. (eds) (2003) *Contested Pasts: The Politics of Memory*. London: Routledge.

Højlund S. (2015) 'Taste as a Social Sense: Rethinking Taste as a Cultural Activity', *Flavour* 4 (6): 1–3.

Holtzman J. D. (2006) 'Food and Memory', *Annual Review of Anthropology* 35: 361–378.

Howes D. (ed) (1991) *The Varieties of Sensory Experience: A Sourcebook in the Anthropology of the Senses*. Toronto: University of Toronto Press.

Howes D. (2003) *Sensual Relations: Engaging the Senses in Culture and Social Theory*. Ann Arbor: University of Michigan Press.

Howes D. (2019) 'Multisensory Anthropology', *Annual Review of Anthropology* 48: 17–28.

Howes D. and Classen C. (2014) *Ways of Sensing: Understanding the Senses in Society*. London: Routledge.

Imbert D. (ed) (2015) *Food and the City: Histories of Food and Cultivation*. Washington, DC: Dumbarton Oaks.

Ingold T. (1996) 'The Past Is a Foreign Country', in Ingold T. (ed) *Key Debates in Anthropology*. New York: Routledge: 163–191.

Ingold T. (2000) *Perception of the Environment: Essays in Livelihood, Dwelling, and Skill*. London: Routledge.

Ingold T. (2008) 'Bindings Against Boundaries: Entanglements of Life in an Open World', *Environment and Planning A* 40 (8): 1796–1810.

Ingold T. (2011) 'Worlds of Sense and Sensing the World: A Response to Sarah Pink and David Howes', *Social Anthropology* 19 (3): 313–317.

Kerner S., Chou C. and Warmind M. (2015) *Commensality: From Everyday Food to Feast*. New York: Bloomsbury.

Klein J. A. (2014) 'Introduction: Cooking, Cuisine and Class and the Anthropology of Food', in Klein J. A. and Murcott A. (eds) *Food Consumption in Global Perspective. Essays in the Anthropology of Food in Honour of Jack Goody*. New York: Palgrave Macmillan: 1–24.

Law L. (2001) 'Home Cooking: Filipino Women and Geographies of the Senses', *Ecumene* 8 (3): 264–283.

Lefebvre H. (1991) *The Production of Space*. Oxford: Basil Blackwell.

Lefebvre H. (1996) 'The Right to the City', in Kofman E. and Lebas E. (eds) *Writings on Cities*. Cambridge: Wiley-Blackwell: 147–158.

Levi-Strauss C. (1983 [1964]) *The Raw and the Cooked: Mythologiques*, Vol. 1, transl. J. Weightman and D. Weightman. Chicago: University of Chicago Press.

Long N. J. and Moore H. L. (eds) (2014) *Sociality: New Directions*. Oxford and New York: Berghahn Books.

MacDougall D. (2005) *The Corporeal Image: Film, Ethnography, and the Senses*. Princeton, NJ: Princeton University Press.

Malinowski B. (1935) *Coral Gardens and Their Magic*. London: George Allen and Unwin Ltd.

Manalansan M. F. (2006) 'Immigrant Lives and the Politics of Olfaction', in Drobnick J. (ed) *The Smell Culture Reader*. London and New York: Bloomsbury: 41–52.

Mann A. M. (2015) 'Tasting in Mundane Practices: Ethnographic Interventions in Social Science Theory', PhD Thesis, Universiteit van Amsterdam, The Netherlands.

Manpreet K. J. (2010) *Transactions in Taste: The Collaborative Lives of Everyday Bengali Foods*. London: Routledge.

Massey D. (1992) 'Politics and Space-Time', *New Left Review* 196: 65–84.

Mauss M. (1990 [1925]) *The Gift*, trans. W. D. Halls. London: Routledge.

Meneley A. (2011) 'Blood, Sweat and Tears in a Bottle of Palestinian Extra-Virgin Olive Oil', *Food Culture and Society* 14 (2): 275–292.

Merleau-Ponty M. (1962) *Phenomenology of Perception*, trans. C. Smith. London: Routledge & Kegan Paul.

Mintz S. (1985) *Sweetness and Power: The Place of Sugar in Modern History*. New York and London: Penguin Books.

Mintz S. (1996) *Tasting Food, Tasting Freedom: Excursions into Eating, Culture, and the Past*. Boston: Beacon Press.

Mintz S. (2005 [1985]) 'Sweetness and Meaning', in Korsmeyer C. (ed) *The Taste Culture Reader: Experiencing Food and Drink*. Oxford: Berg: 110–122.

Nora P. (1989) 'Between Memory and History: Les Lieux de Mémoire', *Representations* 26: 7–24.

Overing J. and Passes A. (2000) *The Anthropology of Love and Anger*. New York: Routledge.

Paddison R. and McCann E. (eds) (2014) *Cities and Social Change: Encounters with Contemporary Urbanism*. London: Sage.

Paterson M., Dodge M. and Mackean S. (2016) 'Introduction: Placing Touch Within Social Theory and Empirical Study', in Patterson M. and Dodge M. (eds) *Touching Space, Placing Touch*. London: Routledge: 1–28.

Petrini C. (2004) *Slow Food: The Case for Taste*. New York and Chichester: Columbia University Press.

Phull S., Wills W. and Dickinson A. (2015) 'Is It a Pleasure to Eat Together? Theoretical Reflections on Conviviality and the Mediterranean Diet', *Sociology Compass* 9 (11): 977–986.

Pink S. (2008) 'Sense and Sustainability: The Case of the Slow City Movement', *Local Environment: The International Journal of Justice and Sustainability* 13 (2): 95–106.

Pink S. (2009) *Doing Sensory Ethnography*. London: Sage.

Pink S. and Howes D. (2010) 'The Future of Sensory Anthropology/the Anthropology of the Senses', *Social Anthropology* (2010) 18 (3): 331–340.

Pitt-Rivers J. (1977) 'The Law of Hospitality', in *The Fate of Shechem or the Politics of Sex: Essays in the Anthropology of the Mediterranean*. Cambridge: Cambridge University Press: 94-112.

Rhys-Taylor A. (2017) *Food and Multiculture: A Sensory Ethnography of East London*. London: Bloomsbury.

Ritzer G. (1993) *The McDonaldization of Society*. Newbury Park, CA: Pine Forge.

Robben A. C. G. M. and Sluka J. A. (eds) (2007) *Ethnographic Fieldwork: An Anthropological Reader*. Malden, MA: Wiley-Blackwell.

Sassen S. (1996) 'Identity in the Global City', in Yaeger P. (ed) *Geography of Identity*. Ann Arbor: University of Michigan Press: 131–151.

Seremetakis C. N. (1993) 'The Memory of the Senses: Historical Perception, Commensal Exchange and Modernity', *Visual Anthropology Review* 9 (2): 2–18.

Seremetakis C. N. (1996) *The Senses Still: Perception and Memory as Material Culture in Modernity*. Chicago: University of Chicago Press.

Simmel G. (1976 [1903]) *The Metropolis and Mental Life*. New York: Free Press.

Siniscalchi V. (2014) 'Slow Food Activism Between Politics and Economy', in Counihan C. and Siniscalchi V. (eds) (2014) *Food Activism: Agency, Democracy, and Economy*. London: Berg: 225–242.

Sobal J. and Nelson M. K. (2003) 'Commensal Eating Patterns: A Community Study', *Appetite* 41 (2): 181–190.

Stoller P. (1997) *Sensuous Scholarship*. Philadelphia: University of Pennsylvania Press.

Strathern M., Peel J., Toren C., Spencer J. and Ingold T. (1990) *The Concept of Society is Theoretically Obsolete*. Manchester: GDAT.

Sutton D. E. (2001) *Remembrance of Repasts: An Anthropology of Food and Memory*. London: Bloomsbury.

Sutton D. E. (2010) 'Food and the Senses', *Annual Review of Anthropology* 39: 209–223.

Sutton D. E. (2013) 'Review of *Transactions in Taste: The Collaborative Lives of Everyday Bengali Foods*, by Manpreet K. Janeja', *Flavour* 2: 22.

Swislocki M. (2009) *Culinary Nostalgia: Regional Food Culture and the Urban Experience in Shanghai*. Stanford, CA: Stanford University Press.

Symons M. (1994) 'Simmel's Gastronomic Sociology: An Overlooked Essay', *Food and Foodways* 5 (4): 333–351.

Taylor L. (1996) 'Iconophobia', *Transition* 69: 64–88.

Thrift N. (2008) *Non-Representational Theory: Space, Politics, Affect*. London: Routledge.

Todorova M. and Gille Z. (eds) (2010) *Post-Communist Nostalgia*. Oxford and New York: Berghahn Books.

Trubek A. B. (2009) *The Taste of Place: A Cultural Journey into Terroir*. Oakland: University of California Press.

Tuan Y-F. (1977) *Space and Place: The Perspective of Experience*. Minneapolis: University of Minnesota Press.

Vannini P., Waskul D. and Gottschalk S. (2012) *The Senses in Self, Society, and Culture: A Sociology of the Senses*. New York and Abingdon: Routledge.

Watson J. L. and Klein J. A. (2016) *The Handbook of Food and Anthropology*. London: Bloomsbury.

Weber M. (1958 [1921]) *The City*. New York: Free Press.

Wise A. and Noble G. (2016) 'Convivialities: An Orientation', *Journal of Intercultural Studies* 37 (5): 423–431.

Zukin S. (2008) 'Consuming Authenticity: From Outposts of Difference to Means of Exclusion', *Cultural Studies* 22 (5): 724–748.

Part I

The city and its other

2 Digging into soil, the senses, and society in Utrecht

Vincent Walstra

It is a sight that summons questions and joy: a hen wallowing in a hole of dirt. Do chickens do this? It seems more like a pig's habit. But the hen enjoys it, and so it makes me smile. I have seen videos of big machines shovelling hundreds of chickens around in overcrowded barns of factory farms. This hen, however, will see no such future for it is walking and playing in the urban garden 'Koningshof' in Utrecht, the Netherlands. Together with her fellow chickens, she shares a chicken coop, and because the coop is always open, her terrain includes the entire farm. The eggs they lay are eaten by Koningshof's gardeners, but the chickens need not worry about being fattened and butchered, as they are living here. I recognise myself in the hen's lightheartedness for it reflects the soothing ambience at Koningshof. Both the hen and I enjoy a spacious environment here, where we dwell each at our own pace. Koningshof offers the hen an environment where she can enjoy sensuous experiences like wallowing in dirt, instead of commodifying the chicken as a machine for producing eggs and meat. Gardeners at Koningshof experience the same. They, too, engage in sensuous interactions with plants, birds, insects, soil, people, and themselves. Why do we see urban agriculture appear, not only in Utrecht, but globally? Why, in a society where preparing food can be cost- and time-efficient, do people prefer to engage with this slow process of food production in their own gardens?

This chapter aims to show how urban gardens in Utrecht enhance the sensorial instead of economic qualities of food. I will argue that the industrial food system, which is built on a rational and economic approach to food and nature, lacks appreciation of sensorial experiences beyond consumption, and hence, the trend of people turning to what I call 'holistic gardening' reveals a societal desire to reconnect with nature and food through the senses. This chapter is based on three months of fieldwork in 2017 with various urban agriculture initiatives in Utrecht, the Netherlands, followed by a year of participation at the Koningshof garden in 2018. My aim in this chapter is to contrast modern standards of food procurement in the Dutch urban environment with new forms of engaging with food and nature. Therefore, besides using qualitative methods of participant observation and interviews to understand the sociocultural phenomenon of urban gardening, I apply autoethnographic elements to deepen the understanding of the societal impact of gardening on urban dwellers. Having been born and bred in the Dutch urban environment, my personal introduction to the ecology of food production proved to be a valuable experience in understanding the societal impact of encounters between the modern urban dweller and holistic agricultural activities.

In the first part of this chapter, I will contextualise the meaning of urban agriculture in Utrecht by giving a historical overview of the meaning behind both *urban* and *agriculture* and the roots of the Koningshof urban garden. In the second section, I will explain how urban gardens confront urbanites with the limited nature of their mainstream worldviews. Thirdly, I aim to show how sensorial engagements within the garden environment enables urbanites to reconsider conventional worldviews by taking on personal experiential knowledge. Finally, I will bring all this together by theorising the societal impact of urban agriculture through its physical presence in the urban landscape.

Towards 'modernity' in urban and agriculture

The separation of food production from the urban landscape can partly be allocated to technological developments enabling mass production and the domination of a capitalist ideology of accumulation, transforming non-urban areas to spaces for agro-industrial production (Harvey 1978; Barthel *et al.* 2015). The shaping of this landscape traces back to the age of Enlightenment that laid the foundations for what we now call modernity. During the Enlightenment and onwards into modernity, Western civilisation developed an ontology in which nature was mastered by humans, passion by reason, and the body by the mind (Harvey 1989). This utilitarian approach to human and natural resources caused a subjection of the physical to the hegemonic idea of growth (Tsing 2013). The growing dominance of the capitalist market accompanied by strategies of commodification and commercialisation in the twentieth century accelerated this process of modernisation. In time, everyday life became subject to constant processes of commodification and the economic valuing of resources and practices, reducing all life to a 'logic of capital' (Rigi 2007: 56).

Since the Industrial Revolution, city populations boomed globally (Smart and Smart 2003), the number of people living in urban areas today exceeding the amount of people in non-urban areas. Michael Hardt and Antonio Negri (2000) explain how industrialisation triggered mass migrations towards factories or harbours as areas of labour concentration, from which many cities grew or emerged. Interestingly, factory work has, for the most part, been outsourced from Western cities to other countries and continents in the twenty-first century (Ong 2006), whilst the urban population in the West keeps on growing. Hardt and Negri (2000) argue that physical labour has become decentred, with the urban space increasingly dominated by a network economy focused on social interaction and characterised by the displacement of production. They explain that where the process of industrialisation resulted in a homogenisation of physical labour along conveyor belts, the shift towards an information society is characterised by the increase of desk jobs behind a computer or at home (Hardt and Negri 2000). What this shift shows is that physical, embodied occupations have been reduced to a minimum while computer-navigated, economically rational work are gaining significance. Increasingly, everyday life is dominated by such economically rational control. This development is central to Annemarie Mol's (2013) argument that contemporary society is dominated by a Western notion of controlling the body's dangerous desires by the rationality of the mind. In this chapter, I argue that sensuous experiences of gardening oppose this ontological standard of modern society. But first, allow me to contextualise the previous analysis to the case study of Koningshof in Utrecht.

In his book about the development of Utrecht's urban food landscape, Frank Stroeken (2012) explains how urbanisation accelerated about 150 years ago. Towards the end of the nineteenth century, the areas surrounding what is now the historical city centre became part of the urban area (Stroeken 2012). This occurred during the Industrial Revolution when the mechanisation of agriculture created a fundamental shift in the Western food supply chain, and the Netherlands specifically (Van Otterloo 2013). From this moment, local agricultural business began to decline in response to a growing preference for large-scale international trade. Together with technological developments in the food industry, a cultural shift redefined the meaning of 'good' food (Mol 2013). New measurements such as hygiene, nutrition, and a long shelf life became important. Anneke H. Van Otterloo acknowledges that food became valued for convenience (2013), placing economic function over sensorial appreciation. The history of the Koningshof urban garden echoes this transition in Western, and specifically Dutch, society.

For over two centuries, the Jongerius family maintained their horticulture farm at the place where we now find Koningshof. On a piece of land of about 13,000 square metres, Robert, Jos, and generations before them cultivated diverse crops to feed people in Utrecht. During our conversations they told me how cultivation traditions were taught by father to son. However, for father Jos and son Robert, things were handed over slightly differently. Jos grew up sitting on the back of his (grand)father's tractor going to the regional market and auction to sell their harvest. He tells me that they used to sell their harvest at what is called the *Jaarbeurs* ('trade fair'). Nowadays the *Jaarbeurs* is known for its big cinema and for hosting events. Its history of food trade, however, is demarcated by the names surrounding the area: the street named *Veemarktplein* ('livestock market square') and the restaurant *Korenbeurs* ('grain exchange'; over the course of writing, the restaurant was renamed) are reminders of earlier days. Jos explains how, in the 1970s, their harvest would determine what people would buy and eat. During his youth, this producer-consumer relationship gradually started to shift. Food supply chains and industrial farming boomed, and small-scale farmers like the Jongerius family increasingly had to compete with low-cost foreign produce that responded to consumer-driven demands for food. From the 1970s onward, the situation worsened, but for a while Jos could stay afloat by changing from horticulture to selling flowers. Eventually, around the turn of the new millennium, he had to quit farming and shift occupations.

The farm lay fallow until 2012, when his son Robert, together with four friends who met during their study of landscape architecture, decided that this land and its identity needed to be revived. They started the foundation *Koningshof* ('Kingsyard', referring to the street it is located on, called *Koningsweg*), aimed at reconnecting the area, local citizens, and anyone interested in the historical identity of the farm, the food, and the region. Their goal was to raise awareness of the unsustainability of the 'modern' food system and to educate people about the meaning of and engagement with natural resources like plants, insects, animals, and the elements. Hardly 30 years ago, growing local food proved to be unfeasible in a globalised food system, but nowadays, there is a waiting list for citizens who want to pay to grow food themselves. What has changed? In the next section, I focus on Koningshof as a place where urbanites are confronted with a lack of gardening skills and become skilled in gardening at the same time. I focus on this transition in particular because it is the principal purpose of Koningshof and puts urban gardening within a wider societal spectrum as opposing modern conventionalities. I will demonstrate the societal value of the Koningshof in breaking through urban and rural, and human and nature, dichotomies.

Beyond dichotomies: holistic gardening

On a Saturday morning, I leave my home near Utrecht's central station to cycle to Koningshof. A ten-minute ride brings me to the city's edge where I park my bike next to the farmer's house and walk towards my garden plot. Passing by the fruit orchard and picnic area, I greet Roeland and Robert, who are standing outside making quiches with freshly picked vegetables for lunch, and Akke, who is in the kitchen inside the greenhouse. Two of my fellow gardeners have already started weeding and hoeing the legumes section of our private 50-square-metre garden, which we have divided into six sections to maintain an organic crop rotation system. As has become our habit, we proudly observe that our sunflower has now grown over a metre tall, even though Matthias accidentally planted it too early in the season. 'That won't last', Jos had assured us. Since then it has become Matthias's and our project and pride. Although we are proud that our sunflower survived our early enthusiastic mistakes, we take Jos's advice to heart. Most Saturdays at the garden involve a chat with Jos, asking for his opinion about our garden, often resulting in a critical analysis of our and others' gardening habits. 'I see people drown plants in water, whilst you should tease them; otherwise they get lazy', Jos explains to us. 'If a plant feels it is dying, it will think of reproduction to maintain its existence. So what do you think will happen? It gets energetic to produce an offspring, which serves us with the parts we want to eat'. Once again, I am astounded by the logic gardening entails. 'Why do you think a carrot grows large? It searches for water deeper in the ground. If you keep the soil moist at the surface, do you think it will have to dig deep to get water?' The logic of plants is so obvious – one might even say natural – that it makes me aware of my lack of understanding of such basic ecological processes.

In his article about 'industrial gardening' in Great Britain, Thorsten Gieser argues that a contemporary inhibition of enskilment due to time pressure is the result of a lack of interest in caring and learning. Today's generation is concerned with 'doing what we are told to do' as quickly as possible so as to have spare time after finishing the task (Gieser 2014: 143), a poignant analysis considering my own research at Koningshof. When I started fieldwork, I simply did what the garden coordinators told me to do and was merely concerned with doing research and much less with caring or learning about gardening. Only when I got my own garden plot after fieldwork did I begin to actually understand the processes of plants, soil, seasons, weather, animals, humans, and the most important aspect: the necessity of harmony between them. The significance of understanding this harmony is explained by a gardener from another urban garden in Utrecht:

> When Mao Zedong was ruler of China he ordered all sparrows to be killed since they ate too much of the grain. His order was followed, and the sparrows extinguished from the land. However, this created an imbalance in the ecology of the land. As a result, insect populations were disturbed and now there is a lack of bees and other pollinators. Consequently, people now have to pollinate orchards by hand.

Symbolic stories like these serve gardeners with a fundamental understanding of the interrelatedness of life processes. To describe the importance of appreciating the process rather than the completion of work, Thorsten Gieser (2014) mentions Tim Ingold (2011). As a recurring theme in his writing, Ingold persists in delivering a fundamental message that seems to grow ever more important: there is no distinction between humans and nature, and life is a continuity, a process without beginning or end (Ingold 2005: 504). In the process of enskilment in the practice of gardening, such ontological lessons are learned through experiences. But this does not mean that all modes of gardening achieve the same outcomes. Gieser distinguishes between 'enthusiastic expert gardeners' and 'industrial gardeners' (2014: 146). The latter are concerned only with finishing a task and have a 'temporal horizon' (ibid). The former cherish and care for plants, developing relationships throughout their engagement. Whereas the idea of fulfilling a task creates a linear mindset, emphasising process instead allows people to experience life as a circular system. It is the latter approach that is experienced in Utrecht's urban gardens. In a linear mindset, harvest as an end-product would matter in the process of food production. But when one of the gardeners tells me her strawberries were eaten by a bird, it does not cause disappointment or anger. Instead, she tells me: 'If the birds ate my strawberries, then that is what they needed to do'. Statements like these are not rare amongst urban gardeners in Utrecht when confronted with the loss of harvest. The egocentric idea of resource extraction is substituted for a holistic understanding of giving and taking between human and nature.

The lack of such a holistic view in contemporary society is demonstrated when Robert quotes a child saying: 'I won't eat that; it was laying on the ground!' On Saturdays, Koningshof is open to visitors to buy food from the garden inside the greenhouse, maintained by the Koningshof initiators. It serves the purpose of letting people walk through the garden and greenhouse and harvest their own products as opposed to shopping from supermarket shelves. By doing so, people get a sense of where the food comes from, how it grows, and who farms it. Hence, through awareness, Koningshof aims to educate people about processes that precede the end products we consume. This quote illustrates the relevance of this engaged shopping as an educational process, especially for younger generations, for this is the response of a child shopping at Koningshof after he realises the lettuce he and his mother harvested will later be served to him at dinner. The encounter confronts the child with the reality of lettuce being a plant, instead of a ready-made product on the supermarket shelf. By facilitating this confrontation, Koningshof contributes to understanding human-nonhuman relationships that relate to food consumption.

In the previous paragraph, I explained how the period of Enlightenment laid the foundations for the modernist interpretation of humans dominating nature. This ontology is built on dichotomies between humans and the world, which, according to Ingold (1993), are unjustified imaginations of reality. This section demonstrated how urban gardens in Utrecht are interacting with the broader nonhuman environment, providing experiences that break through human-nonhuman dichotomies. In the next section, I clarify how holistic gardening is induced by sensorial experiences, integrating the social and ecological and synergising body and mind.

Appreciating sensorial experiences

> I don't know that much about plants, but I learn more here than I could have ever learned from any book.
>
> Martine, urban gardener in Utrecht

Using the theory of social innovation, Jean Hillier (2013) claims that people respond where societal structures fall short to fulfill essential human needs. In other words, innovations are responses to societal *lacks*, which can be traced back to a meshwork of tacit dynamics in society. I argue that the appearance of urban agriculture in the Netherlands is a response to the dominance of economic efficiency over sense in experiencing the everyday. Interestingly, in the English language, 'sense' refers both to embodied perception and reason. It stresses the synthesis of the body and the mind in sensorial experiences. Nancy Scheper-Hughes and Margaret M. Lock (1987) argue that the capitalist ideology dominating Western society has alienated the rational mind from the material body by separating manual and mental labour. In what follows, I will show that urban gardens in Utrecht respond to a lack of balance between mind and body and reconnect manual and mental labour through holistic experiences of the senses.

I am working in the Koningshof greenhouse when I notice Joris digging up plants and putting them in a wheelbarrow. The plants are about half a meter long with many green leaves. On top of the plant there is a purple-white flower not much larger than the tip of my thumb. I don't know the flower, so I ask Joris about it. He explains that the flower is edible, and they use it in salads, after which he picks one and offers it to me. Together we taste the flower. It feels strange to put a flower in my mouth, and I wonder if I have ever consciously eaten a flower before. The structure is soft, the taste is sugar-sweet, almost honey-like, and the aesthetic of the flower affects the experience of eating and tasting its beauty. Joris explains that, as far as he knows, from this plant we can only eat its tiny flower.

Now that I've been made curious, I ask Joris if the flower next to it is edible too. It is a yellow-orange flower looking like a miniature sunflower. At this moment, one of the gardeners, Tom, passes by, and Joris calls him. Tom used to own an organic food store in Utrecht. Without answering my question, Joris asks Tom to tell me about calendula. Enthusiastically, Tom explains the healing power of calendula, which is an ointment or oil used for skin irritation or scratches. When Tom finishes his story about having used it against saddle sore when cycling, Joris adds to the story that the yellow-orange flower is calendula. Tom and his daughter respond with a satisfied nod and proceed to their garden. Then Joris turns to me again and clarifies with a grin: 'Every plant has its own story', after which he continues digging up the plants.

Gardeners like Tom rent their own 50 square metres of land at the Koningshof. But the gardening is far from an individual process. The idea behind this initiative is to provide gardeners with a 'Workshop Koningshof'. This means that from the beginning of the season at the end of March, every Saturday throughout the year, Robert, Joris, Roeland, Gijs, Akke, and Jos, the six who initiated Koningshof, invite gardeners to visit their plots and spend the day together. What they offer the gardeners is a place, material, and knowledge. Place entails both the physical gardens and a community for sharing seeds, food, experiences, and more. The material they offer are tools like spades and wheelbarrows, but also compost, water, and other crucial gardening elements. Finally, knowledge is constantly transferred in short conversations like in the previous example, but also through workshops of tasting and processing food. The purpose of offering these facilities is to create an atmosphere in which people can physically engage with their environment. When I overhear the coordinator of another garden explain to someone the main function of the garden, I hear him say: 'That people feel connected with the earth, soil and the plants. Then awareness will find its way, because once you have eaten something from the garden, you won't want anything different'. Peter, one of the gardeners, confirms the importance of eating from the garden, explaining it as 'going back to the basis of being human'. Besides experiencing taste, physical interaction with the soil often recurs as a sensorial experience that urbanites appreciate about gardening. One of the Koningshof gardeners, reflecting on her occupation as civil servant, distinguishes between 'abstract' work behind a desk and working in 'the physical world'. Chuckling, she adds: 'I like to stand in the clay, literally. I enjoy getting dirty'. In contrast to her work behind a laptop at a desk, the touching of the soil is experienced as more real because of its physical component. The same goes for experiencing the sound of buzzing bees, which does not cause panic and swinging limbs aiming to kill the insects, but rather the opposite: 'They amaze me, they form an essential part of this garden', one of the garden coordinators proclaims when he shows me the gardens' beehives, which house tens of thousands of bees. Similarly, heaps of manure do not raise ugly faces and complaints about stench but instead are spread over the field with care and appreciated for their crucial function as fertiliser. Dirty hands, buzzing insects, nasty smells – they do not belong in the modern urban environment. However, for urban gardeners, they comprise a healthy and desirable environment that modern society has failed to offer them.

The Koningshof initiative offers a place where people learn by doing. The embodied practice in a personal garden enables the gardeners to engage with the process of gardening and develop skills. At the same time, the farm's community, both the initiators and fellow gardeners, share knowledge and practice and hence teach and learn together. In her conceptualisation of enskilment, Cristina Grasseni (2007) uses both an embodied and a social dimension to explain its meaning. Enskilment is embodied in 'material and social learning experiences' (Grasseni 2007: 11), whilst the social dimension of apprenticeship gives practices and experiences a contextual meaning. The integration of both body and mind into skilled practice is emphasised by Ingold, who says that 'skilled practice entails the working of a mind that, as it overflows into body and environment, is endlessly creative' (2018: 159). This process of acquiring knowledge through enskilment is well described in Gísli Pálsson's (1994) ethnography on Icelandic fishers. He explains how body and mind interact in the skill to read the landscape and 'see' (Pálsson 1994: 910) the fish, something Icelandic skippers have learned both through practical engagement with the environment and by working as apprentices with an experienced skipper. Urban gardening in Utrecht encompasses a similar process of enskilment through personal engagement with plants, soil, materials, and elements, whilst farming experience is transferred between people in apprentice-teacher relationships.

For the Jongerius farm, the industrialisation of the food system meant the end of generations of family horticulture. But the place has been revived, albeit in a different form. David Sutton (2001) in his 'anthropology of food' discusses the contemporary deskilment in society when it comes to food-related practices. With technology increasingly replacing human practice, bodily and cognitive skills are being lost. Opposed to cooking with machines like blenders or microwaves is the cooking of food with feeling. In line with examples given by Sutton showing how a 'disdain for technology here goes with a disdain for measurement and precision, seen as part of the alienation of modern life' (Sutton 2001: 133), I argue that urban agriculture is about reengaging with feeling, or rather sensing, the process of food production. In the next and final section, I conclude by arguing that urban agriculture impacts society by altering the urban landscape.

Urban agriculture: reinventing food practices

> The green and tranquility work therapeutically, they create an oasis in the desert.
>
> Rashid, urban gardener in Utrecht

We should not underestimate the societal impact of the simultaneous disappearance of agriculture and food production from everyday urban practices and the constant growth of urban populations. Food is an essential element of life, and with an increasing outsourcing of practices to technology, people are losing the skills to grow and prepare food, consequently losing autonomy to those who wield production machines (Sutton 2001). At the same time, however, its gravity should not be overestimated. Similar to Molly Scott-Cato and Jean Hillier's study of transition towns as spaces of hope and change (2010), the study of urban agriculture is one of opportunity for it demonstrates how society adapts to remedy its fallacies. Urban agriculture in the first place is a 'reinvention' (Grasseni 2013: 40) of food production in the urban space through the (re)adaption of agricultural practices to an urban physical and social environment. Through an 'education of the senses' (Pink 2008: 98), urban agriculture advocates a worldview that counterbalances the modern standard of rational and economic reading of food and nature with oases for sensorial experiences. In line with other research on urban agriculture (Barron 2017; McIvor and Hale 2015; Premat 2009), I argue that urban gardening contributes to a wider cultural movement of contestation and empowerment against dominant ideologies of economic efficiency and the centralisation of knowledge and skill.

Multiple times now I have mentioned the oneness of body and mind, citing various researchers who have claimed the importance of seeing both as inseparable elements of being human. The quote opening this conclusion confirms that the embodied experiences of the urban garden environment soothe the mental state. But it does not only affect the gardeners. By emphasising the physical contrast of the green with the rest of the urban environment, Rashid makes me aware of the wider societal impact urban gardens have. By offering unconventional sensorial engagements in the urban space, these gardens affect how cities are experienced (Pink 2008). In his ethnography of the urban space in Bangkok, Claudio Sopranzetti defines the urban landscape to the ethnographer as a 'canvas' on which one can see the layeredness of urban space (2018: 36–37). Several types of architecture that can be recognised in the urban space of Utrecht reflect stories of different periods in human history. For instance, remnants from the Roman period tell us about the first settlers, whilst the medieval cathedral and parts of the city wall trace back to the beginning of its urban formation. There are typical tiny houses and areas from the nineteenth century, whilst at the same time, entire neighbourhoods of concrete apartment buildings from the 1960s shape the surrounding urban areas. Nowadays in the central station area, shining high-rise office buildings are being built by the dozen. Each of these architectural developments signifies societal transformations. For Rashid, quoted at the beginning of the conclusion, the 'desert' consists of the tall, concrete, and dense structure of the city, combined with a demanding and rushed atmosphere. In contrast, the garden serves as an 'oasis', a converse environment. Gardening works 'therapeutically', generating a sense of mindfulness and place by engaging oneself within the physical environment. This experience symbolises a key function of urban gardens in Utrecht, which now have carved themselves into the urban canvas as sensorial alternative environments, signifying yet another movement in contemporary societal transformations.

References

Barron J. (2017) 'Community Gardening: Cultivating Subjectivities, Space, and Justice', *Local Environment* 22 (9): 1142–1158.

Barthel S., Parker J. and Ernstson H. (2015) 'Food and Green Space in Cities: A Resilience Lens on Gardens and Urban Environmental Movements', *Urban Studies* 52 (7): 1321–1338.

Gieser T. (2014) 'Enskillment Inhibited: "Industrial Gardening" in Britain', *Journal of the Royal Anthropological Institute* 20 (1): 131–149.

Grasseni C. (2007) 'Introduction', in Grasseni C. (ed) *Skilled Visions: Between Apprenticeship and Standards*. Oxford: Berghahn Books: 1–19.

Grasseni C. (2013) *Beyond Alternative Food Networks*. London: Bloomsbury.

Hardt M. and Negri A. (2000) *Empire*. Cambridge, MA: Harvard University Press.

Harvey D. (1978) 'The Urban Process Under Capitalism: A Framework for Analysis', *Readings in Urban Analysis* 2 (1–3): 101–131.

Harvey D. (1989) *The Condition of Postmodernity: An Enquiry into the Origins of Cultural Change*. Oxford: Blackwell Publishers.

Hillier J. (2013) 'Towards a Deleuzean-Inspired Methodology for Social Innovation Research and Practice', in Moulaert F., MacCallum D., Mehmood A. and Hamdouch A. (eds) *The International Handbook on Social Innovation: Collective Action, Social Learning and Transdisciplinary Research*. Cheltenham: Edward Elgar Publishing: 169–180.

Ingold T. (1993) 'The Temporality of the Landscape', *World Archaeology* 25 (2): 152–174.

Ingold T. (2005) 'Epilogue: Towards a Politics of Dwelling', *Conservation and Society* 3 (2): 501–508.

Ingold T. (2011) *Being Alive: Essays on Movement, Knowledge and Description*. New York: Routledge.

Ingold T. (2018) 'Five Questions of Skill', *Cultural Geographies* 25 (1): 159–163.

McIvor D. W. and Hale J. (2015) 'Urban Agriculture and the Prospects for Deep Democracy', *Agriculture and Human Values* 32 (4): 727–741.

Mol A. (2013) 'Mind Your Plate! The Ontonorms of Dutch Dieting', *Social Studies of Science* 43 (3): 379–396.

Ong A. (2006) *Neoliberalism as Exception: Mutations in Citizenship and Sovereignty*. Durham, NC: Duke University Press.

Pálsson G. (1994) 'Enskilment at Sea', *Man* 29 (4): 901–927.

Pink S. (2008) 'Sense and Sustainability: The Case of the Slow City Movement', *Local Environment* 13 (2): 95–106.

Premat A. (2009) 'State Power, Priveate Plots and the Greening of Havana's Urban Agricultural Movement', *City and Society* 21 (1): 28–57.

Rigi J. (2007) 'The War in Chechnya: The Chaotic Mode of Domination, Violence and Bare Life in the Post-Soviet Context', *Critique of Anthropology* 27 (1): 37–62.

Scheper-Hughes N. and Lock M. M. (1987) 'The Mindful Body: A Prolegomenon to Future Work in Medical Anthropology', *Medical Anthropology Quarterly* 1 (1): 6–41.

Scott-Cato M. and Hillier J. (2010) 'How Could We Study Climate-Related Social Innovation? Applying Deleuzean Philosophy to Transition Towns', *Environmental Politics* 19 (6): 869–887.

Smart A. and Smart J. (2003) 'Urbanization and the Global Perspective', *Annual Review of Anthropology* 32 (1): 263–285.

Sopranzetti C. (2018) *Owners of the Map: Motorcycle Taxi Drivers, Mobility and Politics in Bangkok*. Berkeley: University of California Press.

Stroeken F. (2012) *Vlaaien Op de Neude: 2000 Jaar Stadslandbouw En Voedsellandschap in Utrecht*. Utrecht: Terra Incognita.

Sutton D. E. (2001) *Remembrance of Repasts: An Anthropology of Food and Memory*. Oxford: Berg.

Tsing A. (2013) 'Sorting Out Commodities: How Capitalist Value Is Made Through Gifts', *Journal of Ethnographic Theory* 3 (1): 21–43.

Van Otterloo A. H. (2013) 'Healthy, Safe and Sustainable: Consumers and the Public Debate on Food in Europe and the Netherlands Since 1945', in Spaargaren G., Oosterveer P. and Loeber A. (eds) *Food Practices in Transition: Changing Food Consumption, Retail and Production in the Age of Reflexive Modernity*. London: Routledge: 60–85.

3 Food activism and sensuous human activity in Cagliari, Italy

Carole Counihan

This chapter explores the role of the somatic senses in contemporary food activism in the city of Cagliari, Sardinia. Cities have always been a defining force in Italian culture, and Sardinia is no exception. Food historian Massimo Montanari (2017: 17) affirmed, 'Italian identity has been built – and imagined – over the centuries, starting with the cities'. The historical preeminence of cities, he showed, has led to the subordination of the country and farmers, the erasure of their contribution to cuisine, and the devaluation of physical labour. Montanari, however, claimed that recently these attitudes have begun to change, in part due to new approaches to food stimulated by the local food movement, a form of food activism.

Food activism consists of diverse efforts to make food more sustainable, equitable, fresh, nutritious, and tasty (Counihan and Siniscalchi 2014). Between 2011 and 2015 I used ethnographic methods of participant observation and interviews to study food activism in Cagliari. I looked at organic production and distribution; farm to school programs; Slow Food's 'good, clean, and fair food' campaign; an urban garden; a solidarity purchase group; farmers' markets; and more (Counihan 2019). I found that some food activists make long-term commitments; others come and go. Some earn income in the food sector; others are volunteers. But they all share the goal of improving the working conditions, remuneration, and social connectedness of farmers and of promoting the food they produce. One important way that activists further these aims is through engaging the senses and building sensory as well as intellectual identifications with farmers.

This paper focuses on the understudied somatic or 'foundational senses' (Wagenfeld 2009: 48) used in muscle movement and balance and links these to Karl Marx's concept of labour as 'sensuous human activity' (Marx and Engels 1970: 121). Sensory bodily engagement in food activism in Cagliari enables urban consumers to connect holistically with farmers, the land, and food production. I examine somatic actions in gardening, gathering wild herbs in the countryside, visiting farms and production facilities, and working in the fields. I aim to show that this corporeal involvement with food challenges central fractures of contemporary society by reintegrating self and body, consumption and production, and city and country. It enriches identification and empathy with other humans and nature and furthers the goals of food activism for a more just, sustainable, and healthy food system (Counihan and Siniscalchi 2014).

Food activism in Cagliari

Cagliari is the capital of the island region of Sardinia and its cultural, political, commercial, and transportation hub. With a history dating back to the Phoenician settlement in the seventh century BC and a much deeper prehistory, Cagliari has a storied life as an urban centre and a longstanding alimentary symbiosis with the surrounding fertile Campidano region (Casula 2017). Its metropolitan area, with 432,000 inhabitants, comprises over a quarter of Sardinia's total population of 1,648,000 (Tuttitalia 2020a; Tuttitalia 2020b). In spite of a long history of agriculture and pastoralism, today much of people's food is imported from off island and purchased in supermarkets. Cagliari has a high density of supermarkets and over two-fifths of the island's grocery stores (Floris 2010: 43), which stock mostly imported foods (Porcu 2011).

In Cagliari, as in other cities around the globe, urbanites are taking steps to improve the quality, sustainability, and healthiness of their food through a range of oppositional practices (Counihan 2019; Fonte 2013; Grasseni 2013; Hale *et al.* 2011; Jepson 2014; Harper and Afonso 2019). These aim to combat the demographic processes that have characterised not only Sardinia but also much of Italy and Europe since the 1950s – rural depopulation, a precipitous diminution of farmers, and urban growth (Carta *et al.* n.d.). Many Cagliari residents' ancestors moved from the countryside to the city in the inter- and post-war years to escape the relentless labour, meagre pay, and low status of farm work. The population of Cagliari almost tripled from 66,000 in 1921 to 190,000 in 1971 (Tuttitalia 2020c). At the same time, the population employed in agriculture declined sharply. These demographic trends have continued to the present for the same reasons, with an 'epochal internal migration' witnessing one in three Sardinians moving from the interior to the coastal areas and cities, abandoning agriculture and pastoralism for jobs in tourism and the service sector (Carta *et al.* n.d.: 19). But not all rural areas of Italy have suffered decline to the same extent, and those that thrive are near a city and maintain a diversified and multifunctional economy (Rizzo 2016: 249–250). Supporting such a regional food economy and tying it tightly to urban food provisioning is a major focus of Cagliari food activists.

In this chapter I focus specifically on several ways food activists practiced 'sensuous human activity'. To set the stage, I turn to the words of Cagliari organic food shop proprietor Francesca Spiga. Early in the 2000s, she left her job working in the pharmaceutical industry in northern Italy to return home to Sardinia to realise her dream of promoting good, local, sustainable, organic food (Counihan 2014). She partnered with a farming family to sell their fresh

Sardinian produce in her store. But she lamented consumers' ignorance about small-scale farming and their reluctance to support it:

> People buying organic produce have the same attitude as normal people – they want everything free, everything at a low cost. . . . They want to exploit the labour of others and that's it. They make a lot of talk about fair trade, and then don't help their neighbour who is farming and even say it is expensive, they have a whole bunch of issues, they complain about everything. . . . I would like to see them pick string beans, and see what it takes, or strawberries, to work all day gathering strawberries, and how much should they cost? It bothers me. I would resolve these problems by making everyone work a week in the countryside to make them shut up. There is no other way. We have gotten away from reality. . . . You don't see with your own eyes. You don't see reality how it truly is.

Labouring for food would not only enable urban consumers to *see* reality, but also to *sense* it more fully with their active bodies, increasing empathy with and commitment to local farmers.

Sensuous human activity and the somatic senses

My focus on sensuous human activity in food activism grows out of the recent explosion of interest in the senses in food studies and anthropology (for example, Ayora-Diaz 2019; Counihan and Højlund 2018; Howes and Classen 2013; Spackman and Lahne 2019). I have written about how Cagliari food activists use the sense of taste to attract adherents, educate them, and build networks (Counihan 2018). Here, I want to explore a different aspect of sensory experience: the haptic system comprising what Paterson (2009: 768) calls the 'somatic senses', which involve 'feeling the body in movement and action' through coordination, balance, and muscle use (Paterson 2009: 766). The somatic senses include 'kinaesthesia (the sense of movement), proprioception (felt muscular position) and the vestibular system (sense of balance)' (Paterson 2009: 766). Gretchen Reynolds (2019) claims: 'A need and desire to be in motion may have been bred into our DNA before we even became humans and could have helped to guide the evolution of our species'. Food activists engage the somatic senses when using their bodies in walking or physical labour, important forms of sensory experience that enhance self-awareness, health, and identification with other walkers and workers, as Katrín Anna Lund (2005: 28) found in her research on mountaineering in Scotland.

Just as walking engages the haptic system, so, too, does physical labour, encompassed in Karl Marx's concept of 'sensuous human activity', a phrase originally used in his first thesis on Feuerbach (Marx and Engels 1970: 121). In *The Economic and Philosophical Manuscripts of 1844*, sensuous labour is that which 'belongs to the *Sinne*, "senses"' (Struik 1964: 57). Labour, for Marx, is essential to being human – the mode of production is the 'mode of life' (Marx and Engels 1970: 42). Marx recognises that free human labour uses all the senses in harmony with nature but that under capitalism, alienated labour prevails, and workers are distanced from the work process, its products, human essence, nature, and other workers (Marx 1964). Moreover, as Claire White (2015: 70) notes, capitalist labour is characterised by 'sensory deprivation', and 'the individual is estranged from his own body'.

Food activism combats that estrangement by offering opportunities for people to engage in food work. Just as Elizabeth Pérez (2011: 665) finds that kitchen work transmits somatic knowledge fostering Afro-Cuban Santeria religious practice, I suggest that food labour conveys somatic knowledge fostering activism. Moreover, bodily exertion can lead to fatigue and pain, which can enable urban food activists to share in, understand, and appreciate the relentlessly hard work of food producers, as Francesca Spiga emphasised earlier.

Sensuous human activity in Cagliari food activism

My ethnographic research on food activism in Cagliari uncovered diverse initiatives engaging urban consumers' somatic senses. Some initiatives brought food production to the city – such as the urban garden discussed later in this chapter. Others brought city dwellers to the country. For example, the activist women's group *Domusamigas* partnered with renowned geneticist Ceccarelli to plant several fields with diverse strains of wheat to determine the best ones for the Sardinian soil and climate (Counihan 2019: 31). Planting, weeding, and harvesting in the heat and cold immersed participants in nature, tired their bodies, and gave them an experience of farm labour while contributing to scientific efforts to make local agriculture more sustainable. The organic farming cooperatives S'Atra Sardigna and Su Staì both established opportunities for consumers to produce food through what Su Staì called 'your garden at a distance' ('*orto tuo a distanza*') and what S'Atra Sardigna called a 'shared garden' ('*orto condiviso*'), where consumers paid up front for a share of the harvest and could work in the gardens as much, or as little, as they wanted.

In an effort to disseminate Sardinian heritage beans, the Sardinian Regional Agency for Research in Agriculture (AGRIS) organised an educational event in Uta, twelve miles west of Cagliari, in June 2011 for about 35 interested farmers, researchers, and activists (Counihan 2019: 32–35). Inside a conference room, scientists spoke about the beans and showed us diverse examples. Then they took us on a quarter-mile hike out to the bean fields to see the growing plants, walk the rows, smell the soil, hear the insects, feel the heat, and experience fatigue and thirst – engaging our senses in learning about beans and developing commitment to local varieties (Figure 3.1).

The teaching farms program, or *fattorie didattiche* (Counihan 2019), aimed to educate students about food and agriculture and their significance in Sardinian culture. I interviewed farmer Annalisa Lecca about this program in the Medio Campidano province near Cagliari in May 2011. Lecca explained that the curriculum mandated a visit from the farmer to the school, followed by visits from the students to the farm, and then another by the farmer to the school. She

Figure 3.1 Scientist, activist, and farmer in the AGRIS bean fields near Uta

Source: Photograph by Carole Counihan

had planned a lesson around 'his majesty, the Sardinian pig', a heritage breed that she and her husband raised at their farm. Her first visit to the school was a disaster. It was utter chaos, and the students completely ignored her.

Farmer Lecca had to develop a way to get through to her young charges and figured out that 'they wanted to see and touch; they did not want to listen'. So when they visited the farm, she divided them into three groups – one went to the piggery to see the boar and sow and hold the brand new piglets. A second group went outside to a fenced area where they observed and fed the rambunctious adolescent pigs. A third group walked to the garage to learn about the tractor by touching it, sitting on the driver's seat, and listening to the motor. Lastly, they came to know the pigs literally within their bodies by eating a cured pork product similar to prosciutto called *mustela* made from the farm's animals. Farmer Lecca found that marshalling all the senses, including the somatic, enhanced learning for the children, who were enthusiastically engaged in the farm visit, in contrast to their lack of receptiveness in the barren sensory environment of the classroom. This confirmed what one interviewee said: 'I think that approaching food physically is the best way to transmit things'.

Another example of food activism connecting city dwellers with rural producers was the Cagliari GAS – Gruppo d'Acquisto Solidale, or Solidarity Purchase Group. GAS are collections of consumers who buy directly from producers through long-term relationships and commitment to fair prices and organic food (Counihan 2019: 72–79; Fonte 2013; Grasseni 2013). The first GAS was founded in Italy in 1994, and there are between 1,200 and 2,000 groups in 2020, although numbers are hard to estimate because GAS are grassroots endeavours that come and go. The Cagliari GAS was founded in 2010, and by 2015 it had 250 to 300 members, only some of whom were active purchasers in any given week. The GAS provided city dwellers with local vegetables, fruits, cheese, meat, wine, olive oil, honey, eggs, and other products from farmers and shepherds in the nearby countryside. According to longstanding president Lucio Brughitta, its goals were to attain high quality food at fair prices, to sustain the Sardinian economy and culture, and to create solidarity among consumers and producers.

To develop camaraderie and trust, the GAS regularly organised visits for members to its farmer-suppliers to learn about their land, their work, their personal lives, and their challenges. I joined a GAS producer visit and wild herb–gathering expedition in April 2013 with 25 other people. Two GAS members kindly drove my husband and me from Cagliari about twenty miles north to Barrali, where we met up with the other participants and a farmer and forager named Bastiano who was a GAS supplier. Guided by Bastiano, we explored the flora of the beautiful and varied countryside.

We used all our senses as we walked through the unusually lush and damp early spring landscape (Figure 3.2). We looked at far-off panoramic vistas and close-up bushes, grasses, and edible plants. We rubbed leaves between our fingers and smelled them. We tasted wild asparagus, mustard greens, watercress, garlic, celery, several kinds of chard, cardoons, and many other plants (Figure 3.3).

Sharing and learning about Bastiano's foraging, even for just a few hours, was instrumental in getting to know him and developing respect for his knowledge and stamina. Joining with other GAS members was a relaxed and fun way to build solidarity. After walking and using all our senses for a couple of hours, we shared weariness and hunger. We ambled back to Bastiano's cabin where we set up makeshift tables and benches from planks and sawhorses.

Figure 3.2 GAS members on the herb-gathering expedition near Barrali

Source: Photograph by Carole Counihan

Figure 3.3 Forager Bastiano cutting wild cardoons for GAS members to taste

Source: Photograph by Carole Counihan

Figure 3.4 GAS members sharing a meal after the herb-gathering expedition

Source: Photograph by Carole Counihan

We spread out the food everyone had brought, passed it around, and ate together – tasting each other's offerings and experiencing together the bodily sensations of satiety, contentment, and repose (Figure 3.4). After chatting and eating for a couple of hours, we cleaned up, said our goodbyes, and made our way back to our cars and then to the city. Our tired bodies, wind-chafed faces, and grass-stained hands were testimony to our deepened sensory understanding of farming.

Sensuous human activity in the Cagliari urban garden

Producer visits were a good but sporadic way for consumers to experience farm work, whereas the Cagliari urban garden involved an ongoing practice of sensuous human activity in nature. Several studies have revealed the rich somatic engagements and benefits for city dwellers of participating in urban agriculture. In her research on Cittàslow activism, Sarah Pink (2008: 183) found that urban gardens were physical and social places where people could have 'an embodied sensory experience' of working together outdoors, which helped build new relationships and identities. Amy Wagenfeld (2009: 50) found that gardening engaged all the senses – 'touching, body awareness, balance, smelling, seeing, hearing. . . . tasting . . . and movement skills' while people used tools, dug in the dirt, and tended plants – activities which Anne Jepson (2014: 148) found enhanced 'significant sensual aspects of being human'. James Hale *et al.* (2011) found that urban gardens fostered awareness of nature, connection to others, and overall physical and mental health – benefits that emerged in interviews with Cagliari gardeners.

The urban garden was located in a former quarry on a hilltop on the edge of a residential area on the southeastern side of the city near Monte Urpinu park. It started in August 2012 after years of preparation – writing the charter, seeking land and funding, and struggling with the political bureaucracy. Unable to gain access to public property, the garden organisers found a private landowner who granted them use of a former quarry. The garden had 34 plots and 64 members, with many more requests to participate than they had space for. There was no water at the site, so people had to carry it from home, a difficult task that limited how much they could grow.

I conducted interviews in 2015 with Cagliari urban garden president Paolo Erasmo, vice-president Tore Porta, and former member Carla Locci. She described the garden's appeal:

> City-dwellers go crazy for these urban gardens. . . . Those who came to make a garden said to us, 'we like the idea that we can once again do something with our hands'. The idea that they could eat what they grew, the idea that they could do it themselves, and that they could have their own space inside the city, hence in their daily routine, to dedicate to something living.

Urban garden president Paolo Erasmo echoed her sentiments and Francesca Spiga's about the value of manual labour: 'We are so detached from reality that we do not even realise it. This garden brings us again to have our feet on the ground, and precisely with our feet on the ground to get our hands dirty'.

What Erasmo loved most about the urban garden was the ability to walk there, work the land, and experience nature in the city (Figure 3.5):

Figure 3.5 Cagliari urban garden

Source: Photograph by Carole Counihan

> Having been born in the countryside I want to do the same things now in the city where I live that I did as a child, that is to re-appropriate those things that have been denied to me. There is the possibility to do them again, to plant, see trees – why must I do these in the country and not in the city where I live? . . . Without using the car I can leave home with tranquility, walk over, work, and return home.

He continued:

> And then in the evening I can do what I want – go to the theatre, go to the movies. What I mean is to reutilise the things the city offers while also reutilising what the country offers – to marry these two things.

Marrying city and country through sensuous human activity was an important goal and benefit of the urban garden.

Vice president Tore Porta emphasised the particular nature of gardening labour with a neologism; 'we call ourselves *ortigiani*', he said. 'This is close to *orto* [garden] but also to *artigiano* [artisan]. . . . We are artisans of the garden'.

Artisans are skilled workers who demonstrate mastery, are valued for their craft, and have some control over the labour process – fitting the Marxian ideal of sensuous human activity. Tore Porta practiced skilled no-till 'synergistic gardening' or permaculture based on the teachings of Masanobu Fukuoka (2009) and Emilia Hazelip (2003). He led permaculture workshops for children and adults in schools and the community. Because he had lost his salaried job when the Sardinian petrochemical industry collapsed (Piras 2012), gardening was especially important:

> I am unemployed, sixty years old. But I'm not desperate, I am doing this work here that I love, and so I produce my own food and this is really important. Producing your own food, knowing psychologically that you can produce your own food and sustain yourself physically is really important. Then if you do it with a collective spirit, with many other people, it also becomes interior nourishment.

Collectivity made work uplifting, and working together made the collectivity. Porta continued: 'Not for nothing our association is called "Agri-culture: we cultivate relationships" [*Agri-culture: coltiviamo relazioni*]. We really like this because agriculture is a way to create gatherings of people'. To foster community, the urban garden purposely required all members to work together on certain days to plant trees and herbs and to weed and clean the public spaces of the garden. Labouring together fostered social solidarity and 'integration', Erasmo said, and several scholars have confirmed his claim (Hale *et al.* 2011; Harper and Afonso 2019; Jepson 2014).

The gardens engaged not only people's somatic senses in the muscular labour of hoeing, digging, planting, weeding, and hauling water but also their senses of touch, taste, sight, hearing, and smell (Counihan 2019: 40–41). The garden was in a beautiful spot with a panoramic view of the city. The wind blew almost constantly, caressing the skin and muting the urban sounds of car horns, clanking delivery trucks, and motorcycle engines. The *ortigiani* had laid out the garden to look like a flower with individual beds shaped like petals. Aromatic plants lined walkways and released their fragrance to passersby. While we strolled around the garden, Porta pointed out edible cultivars and gave us several herbs to taste (Figure 3.6). He explained that they had given attention to the 'aesthetic' aspect of the garden because 'if what you see pleases you, you are happy inside'.

Deployment of the body in skilled non-exploitative garden labour generated that happy feeling inside and illustrated the concept of work as sensuous human activity. It recalled anthropologist Sidney Mintz's (2011) discussion of a Puerto Rican ditch-digger or *palero* and Jane Collins' (2011) response to Mintz. Mintz acclaimed the *palero*'s strong, astute, and rhythmic motions digging ditches, which Collins (2011: 436) defined as 'skill – a sensuous reality of attuned perception, pattern recognition, and deft movement'. Although such

Figure 3.6 Urban garden vice president Tore Porta tasting wild plants in the garden

Source: Photograph by Carole Counihan

skilled physical labour has become increasingly rare under industrial capitalism, it can appear at its margins: for example, in food activist enterprises. These can contribute to 'a larger sense of being someone who matters, who lives in history, and who can make change' (Collins 2011: 438). This narrative of agency emerged in the sensuous human activity performed by food activists. Urban garden president Paolo Erasmo articulated it when he said the garden's aim was 'to keep history alive where possible, the important things representing the work of the people'.

Conclusion

I hope to have shown that attention to sensuous human activity can enhance food activism and research about it. Activists walk and labour for food in diverse contexts – in urban gardens, visits to producers, and hands-on educational events. Some, like urban garden president Erasmo and vice president Porta, make long-term commitments and contribute many hours every week; others participate occasionally in a farm visit, tasting, or work party. For all of them, the deployment of the somatic senses has beneficial effects on health and wellbeing (Jepson 2014) and plays an important role in engaging activists in alimentary system change, specifically by creating identifications with producers and connections with land and food.

It is important to acknowledge that walking and labouring for food play a very small part in most food activists' lives and that they practice multiple strategies, including ethical consumption, political organising, educating, and allying with like-minded associations (Counihan and Siniscalchi 2014). But employing the somatic senses is important and seems to tap into a need. In 2018 the Italian Coldiretti farmers' union reported that six out of ten Italians grew herbs, flowers, and vegetables on terraces, on balconies, or in gardens (Coldiretti 2018), showing people's desire to use their bodies and 'get their hands in the dirt'. Just as Pérez (2011: 673) found that newcomers to Santeria who do more sensory labour in the kitchen are more likely to pledge to the religion, perhaps the more food activists use their somatic senses, the more likely they are to maintain a long-term commitment to food system change.

One thing people learn by tending a garden, feeding pigs, or foraging for wild greens is that agricultural labour is hard. Experiencing the exhausting work of stooping, digging, planting, and weeding in heat or cold enables urbanites to empathise with and value the work of farmers. But in the Cagliari urban garden, having to carry water on top of all the other labour was a serious burden and eventually led to the demise of the garden in October 2016. This pointed to the conundrum of human sensory activity through labour – how much is too much? How much is not enough?

The somatic activities practiced by food activists, even assiduous urban gardening, hardly approach the gruelling work of full-time farmers or agricultural labourers. Because activists' work is voluntary and self-directed, it is socially valued. Mintz (2011: 418) emphasised that people take pleasure in work, even if it is exploitative and physically difficult, if it has social value. But historian Montanari (2017) has shown that farmers in Italy have long been demeaned as backward and ignorant. He believes, however, and I hope to have corroborated, that farmers have recently gained status due in part to food activism. I contend that events enhancing activists' sensory engagement in food labour can increase respect for farmers and connect urbanites and rural dwellers, consumers and producers, city and country. It can also contribute to the ongoing re-evaluation of farm work as admirable rather than lowly and a redefinition of the farmer as resourceful entrepreneur rather than ignorant peasant. The more sensory labour and knowledge activists share with each other and with producers, the more they are likely to share identity, form community, and continue working for a more just and sustainable food system.

References

Ayora-Diaz S. I. (ed) (2019) *Taste, Politics, and Identities in Mexican Food*. Oxford: Bloomsbury.

Carta V., Lobina E. and Muscas F. (nd [2013?]) *Dinamiche e tendenze dello spopolamento in Sardegna – Focus sulle aree LEADER*. Cagliari: Programmi di sviluppo rurale Sardegna, Regione autonoma della Sardegna, www.regione.sardegna.it/documenti/1_26_2010 1116090620.pdf.

Casula F. C. (2017) *Breve Storia di Sardegna*. Sassari: Carlo Delfino Editore.

Coldiretti. (2018) 'Caldo, sei italiani su dieci al lavoro in orti e terrazzi', 27 May, www.coldiretti.it/economia/caldo-italiani-10-al-lavoro-orti-terrazzi accessed 24 April 2019.

Collins J. (2011) 'Narratives of Skill and Meaning Within "Menial" Work', *Review (Fernand Braudel Center)* 34 (4): 435–438.

Counihan C. (2014) 'Women, Gender, and Agency in Italian Food Activism', in Counihan C. and Siniscalchi V. (eds) *Food Activism: Agency, Democracy, Economy*. Oxford: Bloomsbury: 61–76.

Counihan C. (2018) 'Taste Activism in Urban Sardinia, Italy', in Counihan C. and Højlund S. (eds) *Making Taste Public: Ethnographies of Food and the Senses*. Oxford: Bloomsbury: 155–167.

Counihan C. (2019) *Italian Food Activism in Urban Sardinia: Taste, Place, and Community*. Oxford: Bloomsbury.

Counihan C. and Højlund S. (eds) (2018) *Making Taste Public: Ethnographies of Food and the Senses*. Oxford: Bloomsbury.

Counihan C. and Siniscalchi V. (eds) (2014) *Food Activism: Agency, Democracy, Economy*. Oxford: Bloomsbury.

Floris F. (ed) (2010) *L'agricoltura nella Sardegna in cifre 2009*. Rome: INEA.

Fonte M. (2013) 'Food Consumption as Social Practice: Solidarity Purchasing Groups in Rome, Italy', *Journal of Rural Studies* 32: 230–239.

Fukuoka M. (2009) *The One-Straw Revolution: An Introduction to Natural Farming*. New York: NYRB Classics.

Grasseni C. (2013) *Beyond Alternative Food Networks: Italy's Solidarity Purchase Groups*, Oxford: Bloomsbury.

Hale J., Knapp C., Bardwell L., Buchenau M., Marshall J., Sancar F. and Litt J. S. (2011) 'Connecting Food Environments and Health Through the Relational Nature of Aesthetics: Gaining Insight Through the Community Gardening Experience', *Social Science and Medicine* 72: 1853–1863.

Hazelip E. (2003) 'Synergistic Agriculture and Permaculture', *Permaculture Activist* 50: 9–10.

Harper K. and Afonso A. I. (2019) 'Food Values in a Lisbon Urban Garden: Between *Sabor, Saber*, and the Market', in Harper K. and Siniscalchi V. (eds) *Food Values*. Oxford: Bloomsbury.

Howes D. and Classen C. (2013) *Ways of Sensing*. Florence, England: Routledge.

Jepson A. (2014) 'Gardening and Wellbeing: A View from the Ground', in Hallam E. and Ingold T. (eds) *Making and Growing: Anthropological Studies of Organisms and Artefacts*. Surrey, England: Routledge: 147–162.

Lund K. (2005) 'Seeing in Motion and the Touching Eye: Walking Over Scotland's Mountains', *Etnofoor* 181: 27–42.

Marx K. (1964) *The Economic and Philosophical Manuscripts of 1844*, Struik D. J. (ed), Milligan, M. (trans). New York: International Publishers.

Marx K. and Engels F. (1970) *The German Ideology*, Arthur C. J. (ed). New York: International Publishers.
Mintz S. W. (2011) 'Caribbean History, Caribbean Labour', *Review (Fernand Braudel Center)* 34 (4): 407–419.
Montanari M. (2017) 'And at Last, the Farmers Win', in Naccarato P., Nowak Z. and Eckert E. (eds) *Representing Italy Through Food*. London: Bloomsbury: 17–32.
Paterson M. (2009) 'Haptic Geographies: Ethnography, Haptic Knowledges, and Sensuous Dispositions', *Progress in Human Geography* 33 (6): 766–788.
Pérez E. (2011) 'Cooking for the Gods: Sensuous Ethnography, Sensory Knowledge, and the Kitchen in Lucumí Tradition', *Religion* 41 (4): 665–683.
Pink S. (2008) 'Rethinking Contemporary Activism: From Community to Emplaced Sociality', *Ethnos* 73 (2): 163–188.
Piras P. (2012) 'Italia e Sardegna: un caso di colonialismo industriale', *Paginauno* 30, December 2012–January 2013, www.rivistapaginauno.it/alcoa-carbosulcis-sardegna.php.
Porcu P. (2011) 'Spesa alimentare nell'isola, metà incasso vola in Europa', *La Nuova Sardegna*, April 21.
Reynolds G. (2018) 'To Move Is to Thrive: It's in Our Genes', *New York Times*, 15 May 2019.
Rizzo A. (2016) 'Declining, Transition and Slow Rural Territories in Southern Italy: Characterizing the Intra-Rural Divides', *European Planning Studies* 24 (2): 231–253.
Spackman C. and Lahne J. (eds) (2019) 'Sensory Labour: Considering the Work of Taste in the Food System', special issue of *Food, Culture and Society* 22: 2.
Struik D. J. (1964) 'Introduction', in Marx K. and Struik D. J.(eds) *The Economic and Philosophical Manuscripts of 1844*, Milligan M. (trans). New York: International Publishers.
Tuttitalia (2020a) 'Città Metropolitana di Cagliari', www.tuttitalia.it/sardegna/provincia-di-cagliari/.
Tuttitalia (2020b) 'Regione Sardegna/*Sardigna*', www.tuttitalia.it/sardegna/.
Tuttitalia (2020c) 'Censimenti popolazione Cagliari 1861–2011', www.tuttitalia.it/sardegna/89-cagliari/statistiche/censimenti-popolazione/.
Wagenfeld A. (2009) 'It's More Than Seeing Green: Exploring the Senses Through Gardening', *Journal of Therapeutic Horticulture* 19: 46–52.
White C. (2015) 'Sensuous Communism: Sand with Marx', *Comparative Literature* 67 (1): 62–78.

4 Humming along

Heightening the senses between urban honeybees and humans

Ferne Edwards

The local food movement has emerged in recent decades in defiance against industrial agriculture and climate change where people are responding to environmental, social, and economic issues by taking back control of their food sources (Cockrall-King 2012; Goodman *et al.* 2011). Urban beekeeping is part of this sustainable food movement. Spurred on by global threats to bee populations, there has been an escalation in beekeeping as a hobby. Recognising cities as places of both refuge and conflict, new tools and understandings are required to navigate increased urban human/nonhuman proximities. However, while much scholarship focuses on peoples' desire to reconnect to their food supply and relocating this production to the city, little literature has explored the role of the senses in guiding this shift. This chapter argues that the senses offer an important source of knowledge that can help ease tensions between the binaries of city/country, consumption/production, and human/nonhuman, offering a pathway to reconnect to place and to human and nonhuman 'others'. By suggesting that we learn to listen to bees and to each other, this chapter advocates for a more-than-human methodology of the senses in which pace, attentiveness, and enskilment can strengthen skills towards creating convivial, multispecies cites.

Up until the twentieth century, Western cities were full of working animals that provided transport, machinery, food, fuel, manure, and a means of trash disposal (Atkins 2012; Blecha 2007; Brinkley and Vitiello 2014). However, changing perceptions of hygiene and a shift in moral values and class politics led to a banishing of many productive animals from urban centres (Philo and Wilbert 2000). This progressive dearth of animal presence culminated in the recent 'insect Armageddon' (Carrington 2019; McKie 2018), resulting in fewer urban insects due to pollution, habitat loss, pesticide use, and global warming.

In recent years, advocates of the local food movement have been seeking 'to bring nature back' not only by providing food but also through their presence and practices to produce environmental and social co-benefits (Cockrall-King 2012; Mata *et al.* 2019; Wolch and Emel 1995). A wave of backyard beekeepers emerged as part of this movement in the early 2000s, attributing values of local, quality, and ethics to food production. For bees, cities provide a stable climate with diverse resources, where people can benefit from bees' presence through pollination, honey production, and as a popular pastime. Urban bee populations soared globally as aficionados kept small numbers of hives in a growing number of locations, increasing encounters between humans and honeybees. At the same time, increased human/nonhuman proximities intensified a human fear of bee stings and swarms, creating conflict with bees and their keepers (Edwards and Dixon 2016). The senses of both beekeepers and citizens play a crucial role in identifying, preventing, and negotiating such threats. Beekeepers rely on the senses to motivate their engagement in beekeeping, to approach and manage the hive, and to detect anomalies in hive behaviour and health. Moreover, as we learn from the interviews reported later in this chapter, beekeepers are also trying to awaken the senses of citizens to the benefits of beekeeping and urban nature.

This chapter draws on Allison and Jessica Hayes-Conroy's (2010) framework of 'mattering, relating and defying [boundaries]' to apply to the practice of urban beekeeping. Hayes-Conroy and Hayes-Conroy interrogate 'the visceral': 1) in their understanding of the agency of physical matter; 2) in moving beyond the individual towards more contextualised, interactive, and articulated understandings of selves and others; and 3) by questioning boundaries to re-imagine how '[the political] lives in, through, and beyond dualistic tensions' (2010: 1276). They argue that engaging the senses is a useful approach to 'feel one's way' into nonhuman worlds, displacing human-centricity to consider, respond, and adapt to the needs of the nonhuman 'other' through listening, touch, smell, and sight. The subheadings – 'sparking' and 'reading' the senses and sensing 'in' and 'across' – step the reader through forms of human/nonhuman engagement in urban beekeeping.

The research draws on qualitative interviews conducted with backyard beekeepers in Australian cities. I conducted 60 interviews with beekeepers and associated stakeholders, and as a participant, I observed related events. This research was conducted in two stages: initially in Sydney in 2016 as part of a national study on the impact of climate change on health and peri-urban agriculture, to develop into an independent project conducted in Melbourne and other parts of Australia.

In Australia, a wide variety of beekeeping 'types' exist, including commercial, 'side-liners' (part-time commercial beekeepers), and hobbyists who keep bees on different scales; use different approaches, hive models and bee species; and embrace different motivations. So, too, in Australia does urban beekeeping represent a transitional space where traditional hobby beekeepers, often male retirees, coexist and share their beekeeping knowledge with often-younger beekeepers of both genders who are typically motivated by environmental concerns (Edwards and Dixon 2016). This chapter demonstrates how beekeepers apply affective strategies to make 'sense across' species, places, and people. However, recent shifts within urban beekeeping practices can overlook this aspect while the role of the senses remain largely lacking within literature on food and cities. This paper argues for the acknowledgement of the senses within these fields to contribute to richer understandings of and methods for working towards the possible coexistence of human and nonhuman urban worlds.

The role of the senses in beekeeping

Sparking the senses

Beekeeping is 'mattered' through the senses in common first encounters such as the smell and taste of raw, fresh honeycomb or from the pain of a first sting, where memories linger to later catalyse into action when opportunities arise to become a beekeeper. One respondent spoke of his father, who, at sixteen:

> Drove past a beekeeper extracting [honey] on the side of the road. . . . He can still to this day, and he's 80, remember the smell of the honey coming out of the van. . . . He's always had bees around him [since].

Another respondent, himself in his 60s, remembers how:

> My grandfather many years ago used to keep bees somewhere in god-forsaken Central Europe. And I still remember the smell and taste of his honey which was something that I could not replicate until recently.

Such memories persist across cultures, lifetimes, and generations. Opportunities to act on these memories often appear at retirement age when elderly men are looking for a pastime. This interest is often piqued by the vital agency of the hive – epitomised by Jane Bennett's (2004) 'thing-power', in which wilfulness and recalcitrance emanate from the vibrant materialism of lively matter – that hooks their curiosity to engage and sustain their involvement. An example of bees' vital presence was witnessed during a beekeeping workshop by an attendee:

> Within that two hours the sound from that hive, from that one frame of bees, goes up and down audibly and . . . the people watching heard this noise and they didn't know where it's coming from. . . . [I]t's like an 'up' in pitch and I looked inside and they were all vibrating their wings. . . . [T]he woman was looking around and thought that the air conditioning had come on or that there was street-sweeper outside or something! They couldn't work out where it was coming from! . . . The bees [had] just got maybe more agitated for a while, and then they all send a signal to each other and then their pitch just changes.

This unnerving power of nature to unsettle the attendees highlights how detached people have become from the physical 'wild' environment where, when even seeking out to understand nature in a workshop, they are unable to recognise its signs.

Reading the senses

Once hooked, beekeepers quickly learn how the hive's vibrant matter 'pushes back', recognising that the colony will alter their behaviour on their own terms, where a beekeeper's ability to manage their bees will be an ongoing challenge that can never be fully mastered. An elderly hobbyist describes this realisation:

> You've got this group of stinging insects that are doing their own thing and you've got to adapt and work around their behaviours to get them to do more or less what you want them to do.

For beekeepers, 'failure' – such as the death or abscondment of a hive – is never an end in itself but instead a step towards accruing greater expertise in which the senses can guide these experiences. Pieta Hyvärinen (2019: 375) recognises that:

> Becoming an expert beekeeper therefore entails developing ways of knowing which not only value the diversity and changeability of practices and knowledges, but also engender openness towards surprises, even the troublesome ones.

So, too, is this knowledge holistic: a beekeeper must learn to read the surrounding environment to ask: 'What is in bloom? Is there enough resources? Is the colony preparing to swarm?' A responsible beekeeper must learn to read the subtle changes that vary in space and season as no two years or places are ever the same. Within an urban environment, the need for this holistic awareness is even greater as beekeepers need to be sensitive to proximate human and urban needs. Beekeeping thus lures people in, while its complexity entices them to stay.

Such dynamism of multiple factors implies a need to fully 'dwell' in both place and moment (Ingold 2000): to feel, observe, and understand complex environmental variables, including elements of pests, disease, the colony's free will, and weather factors. Such 'situated learning' approaches a nearing intimacy of 'being in the know' and, ultimately, of power (Lave and Wenger 1991), where a beekeeper can learn to be responsive and adaptable to changing contexts. A 'good beekeeper' hence needs to acquire a sensual knowledge that is based in a particular place (Hyvärinen 2019).

The senses add an essential layer to understanding this data. They tell us that we cannot foretell, control, or treat 'nature' – or even a single species or hive – as one and the same. Beekeepers will often describe the temperaments of different hives where, sitting side by side, one cranky queen has influenced the colony and 'a smoker should be used', requiring special treatment to calm the bees down. By taking the senses into account, the minutiae in diversity open up. The senses of sound and smell are particularly telling of the state of a hive.

Sound and smell are key indicators of a colony's health and mood. A bee's 'buzzing' can come from flight, as communication amongst the colony, or as a sign of distress. A hive 'roars' when it is queenless while silence may indicate death or abandonment of the hive. Different smells, a beekeeper explains, can convey an environmental lack, disease, or stress: 'If you get that sweet smell you know that there's a nectar flow going on'. Other scents reveal disease. For example, the disease American foulbrood smells like fermenting sauerkraut, while distress reeks of bananas:

> They let off a pheromone that smells exactly like bananas. And it's quite full on and quite nice but you also know that. Jeez! They're really suffering at the moment! They're really, really upset!

Another beekeeper comments that she 'can always pick where a feral hive is because I can smell it'. Hive inspections, an important regular beekeeping activity, are thus multi-sensory: good sight is essential to count the brood, by lifting frames you can feel their weight, and smell and sound convey the health of the hive (Moore and Kosut 2013).

Sensing in

Rather than merely reading the signs, a beekeeper must also adapt their behaviour in order to conduct hive inspections for pests, disease, and potential swarming. Here, the dominance of humans over animals is inverted, where 'the human body is relational' (Hayes-Conroy and Hayes-Conroy 2010: 1277) to the nonhuman, as bees determine what people wear and how people move, smell, and even feel. In other words, bees' 'thing-power' comes to the fore in close proximity, gaining power to induce a physical and emotional response in others. Bennett describes thing-power as 'the lively energy and/or resistant pressure that issues from one material assemblage and is received by others' (Bennett 2004: 365). Thing-power acknowledges relationships between human and nonhuman 'things' and their flowing impacts on the other. Thing-power 'entails the ability to shift or vibrate between different states of being, to go from trash/inanimate/resting to treasure/animate/alert' (Bennett 2004: 354).

For beekeepers, a 'sensing in' occurs when, rather than only donning a bee suit and veil with trousers tucked into long socks with thick gloves and smoker in hand, they must also consciously slow down, breathe, move calmly, and be extremely attentive. Within those few metres, they are entering the hive's domain, and if they deviate from this behaviour, they will suffer consequences. A young mother beekeeper reflects on her experiences when human time pressures disrespected nature's pace:

> I once looked in the beehive when I was rushing and didn't use any smoke and that was a hideous experience for me and bees. And they got very angry! I got agitated and we were like feeding off each other. And a lot

> would have died because they were stinging me. And I carried on because I felt like I had to finish what I was doing! . . . But I will never do that again. Because they remained agitated for about five days afterwards. And they came [and] they were pinging on the back door to tell us, you know, 'we're really unhappy!'

This enforced behaviour from bees to humans also induces a mindful state, as explained by one beekeeper who finds it to be one of her favourite things:

> I think it's quite meditative. . . . You have to be so conscious of what you're doing when you're doing it. It's not just aimless. . . . Say if you're gardening, your mind wanders off and does other things. But certainly not when I'm looking in the beehive. You have to focus totally on what you're doing and relax but you've got to be consistent.

When done well, this resonance across species can be described as a calm 'hum' between the actions of the beekeeper and the tolerance of the hive. At this moment, the beekeeper 'dwells' in their immediate environment (Ingold 2000), keenly aware of the movements and atmosphere around them. This shift in pace and focus can lead to a 'more-than-human' experience, during which some beekeepers lose track of the boundaries of themselves, eliciting a form of detached introspectiveness. This experience is described by a retiree beekeeper (cited in Edwards and Dixon 2016: 546):

> You, I suppose, feel a little bit small because . . . you become part of nature without really admitting too much about it. . . . It seems soft and unmanly sort of thing, but it does give you a clear . . . You learn to appreciate the cycles of life and death. . . . And you're watching it and you think to yourself, well my hives died, well so am I going to die too and so is everybody else around me going to die so therefore you don't. . . . You sort of accept it a lot easier. It helps.

For commercial beekeepers who spend days with their hives as they follow the honey flow across the Australian countryside, this sense of nature connection can be even more piquant. When asked to describe his favourite experience, a commercial beekeeper replied:

> We were getting ready for the next day [when] we're going to pull honey off them. But we're on the back of the truck in the swags. Looking up. It was pitch black. The Milky Way – you felt if you put your hand up, you could just grab hold of it. It was that bright above you. And the noise of the bees as they were curing nectar would be one of the most relaxing moments I've had in my life. And that's really – those sort of little moments – in what we do is just phenomenal. And for an urban beekeeper I actually pity them that they can't actually experience it that way because

> where we were parked we were literally – there's no lights, there's no cars, there's no people. We're just stuck in the middle of nowhere. And you've got this going on beside you and you're just lying there looking up and seeing stars so bright.

Hence, beekeepers become 'attuned' (Ingold 2000) to nonhuman worlds as they use their senses to open themselves up, slow themselves down, and change their behaviour to learn to listen both to nature and to themselves. Urban beekeeping, through its complexity of care and focused attention, is similar to (yet I argue goes beyond) Russell Hitchings's description of gardening that draws humans 'down into their world, and make for an understanding of their concerns and a commitment to their care' (Hitchings 2003: 107). This shift is akin to Donna Haraway's 'becoming with' across species, when, in her words, 'Once "we" have met, we can never be "the same" again' (Haraway 2008: 287). Such continuous processes of entanglement between human and nonhuman worlds also evoke notions of 'being with' for Nick Bingham (2006), 'being-for-the-other' for Paul Cloke and Owain Jones (2003: 210), and as 'livingwith' for Maria Puig de la Bellacasa (2017: 83). Alternatively, Hyvärinen (2019) expresses 'diverse and particular practices of securing the necessities of life through sharing lifeworlds with human and non-human others' as 'multispecies livelihoods'. These iterations highlight the various dimensions that caring relations can take across human and nonhuman domains, where 'such thinking is a form of 'ethical mindfulness' that decentres and deprivileges humans in multispecies relations, and engenders respect for non-human species' (Maller 2018: 121).

Sensing across

The use of the senses to bridge human and nonhuman worlds also extends to smoothing differences across people from different backgrounds. To 'sense across' further builds on situated knowledge, to consider how one learns through the experience of sensing in 'doing'. Noting the wide diversity of beekeepers in Australia mentioned earlier, clashes could occur between beekeepers due to differences in approach, purpose, background, and experience. One such clash involves the different types of beekeeping training and knowledge experienced by commercial and hobbyist beekeepers, where the former are often taught in the field over long periods of time, receiving little formal education. The author interviewed a commercial (middle-aged) beekeeper who recounts a conversation with an amateur. He first explains:

> We're happy to talk to anybody about beekeeping. An amateur comes to us, we're happy. Because it's a benefit to us to actually help educate them and talk to them to highlight some of the things that we see with bees. Because you know if they don't understand disease and the problems, it impacts us.

This quote exemplifies how beekeeping is a community of practice, where the actions of one impact the welfare of many (Pálsson 1994). To ensure a safe environment for bees, all beekeepers – no matter what their background, approach, or purpose – must responsibly treat pests and disease. In the compressed confines of the city, beekeepers take on more responsibilities that can impact the beekeeping community at large, where they must also consider the concerns of their non-beekeeping neighbours. Thus, ways to improve understanding across beekeepers, and between beekeepers and the public, are paramount to safeguard both bee health and the practice of urban beekeeping overall.

This conversation described in the interview between the commercial and novice hobby beekeeper takes a downturn when the commercial beekeeper identifies a lack of appreciation for experiential knowledge. The commercial beekeeper recollects the novice remarking, 'Well, I've watched a lot of things on YouTube. And I've read everything going on that, I've googled [what] I can. And I know beekeeping'. In the interview, the experienced beekeeper burst into laughter. He explains to the author:

> I can point out a lot of [commercial, rural] beekeepers who can't even read and write. . . . And hold on a sec! We can pull a hive apart and tell you by looking at it or by listening to it, what's wrong with it, and what you can do to fix it. So we mightn't have the skillset to write a PhD about it, but we have the, you know, we have the 'hands on' knowledge of what to do.

Here, the commercial beekeeper highlights the value of experiential knowledge that is conferred through the senses – as described in 'reading the senses' earlier. In the commercial beekeeper's view, an appreciation for long-term, situated knowledge is displaced by placeless, emerging technologies that, in turn, devalue the long-term knowledge – in this case, that of commercial beekeepers. Experiential knowledge – knowledge by doing, accumulated through years of practice – is crucial for beekeeping, where the senses both guide and provide shared understanding across people of different backgrounds and across species, combined with the slowing down and reflection of situated learning to better understand bees. Gísli Pálsson (1994: 901) defines enskilment as a way of learning by 'immersion in the practical world'. The value of this knowledge is swiftly displaced by placeless methods, such as online media. While technology can provide the descriptions of beekeeping technique, it remains bereft of both the senses and the social context in which it is practiced. Hence, even sense as new knowledge is not itself a private matter but emerges from participants' histories and their socialisation within communities of practice that exists around them (Barth 2002).

An often-successful example of shared learning in beekeeping is informal (unpaid) teacher-apprentice mentorships (Lave 2011). Mentoring occurs within (often informal) beekeeping clubs and as a 'buddy' system in the field. Exchanging knowledge over time, both approaches develop personal relationships with the common interest of bees forming a base to relate and listen to each other. Recognising that many older male beekeepers (who take the teacher role) are quite shy, this one-to-one relationship allows people to connect with others in spaces they feel comfortable, where people can ask questions and be repeatedly shown useful techniques. Mentoring is expected to be reciprocal; newcomers who receive advice provide 'a second pair of hands' to the established beekeeper and may, in time, become mentors to others themselves. With no formal training programs required in Australia, and due to the complexity of beekeeping, mentoring provides a 'thinking-through-doing' approach that is based on visceral, practical, and grounded experience. A novice hobby beekeeper explains:

> Book learning is one thing. And it affords a certain amount of knowledge about a subject but when you have somebody who can guide you through the experience and talk to you on a human level about what's actually occurring in your endeavour, I think that helps you understand better. It helps me understand better. I find that to be a far more intuitive way of learning and I think a more effective way of learning. . . . Especially for something like beekeeping, it's not a science per se, it's not a mathematical equation where there's only a couple of ways to solve the problem. You know, it's an organic, natural endeavour and there's lot of different ways to do things and everyone has an opinion about it. And there's not a great deal of science around the right way to keep bees. So it's really important to speak to lots of different people and learn from their experiences. And I say 'experiences' because everyone has a different experience of beekeeping. And then to decide what the best way to do it for you is in your situation.

In addition to dwelling with bees in mediation and place, beekeeping through mentorship represents 'a process that allows us to attain richer and fuller translations of bodily experience and materiality that are located, multi-textured, reflexive, sensory and polysemious' (Witmore 2004: 60). Such varied, hands-on, shared, and lived connections enable beekeepers to make sense of their bees and each other and are vital for establishing solidarity in an increasingly diversified beekeeping community.

Furthermore, beekeepers' participation in the wider community by holding open days at local beekeeping clubs and community gardens and by participating in events such as school visits allows them to share the intricacies of beekeeping with the public through the senses to help overcome their fear. The conversion of the public's 'fear' of urban bees into one of 'care' could have far-reaching consequences, such as promoting conservation efforts. However, such efforts need to be further mobilised and examined as, while beekeeping numbers have increased, fear of bees still tends to predominate in the wider public in many cities.

Making sense of

Acknowledging that beekeeping stirs an affective shift, how do shifts in affective states impact political a/effects? (Goodman 2015: 5). Bennett recognises that the political potential of vibrant matter lies 'in its ability to induce a greater sense of interconnectedness between humanity and non-humanity' (Bennett 2004: 367). So, too, do Hayes-Conroy and Hayes-Conroy in 'defying boundaries' bring in politics to discourage dualisms – between mind and body and city and country – to enable a 're-imagining and practicing of (political) lives in, through, and beyond such tensions' (Hayes-Conroy and Hayes-Conroy 2010: 1274). Hayes-Conroy and Hayes-Conroy acknowledge that: 'by paying attention to how matter is mobilized in these ways – how different bodies are moved to do or act – we can begin to recognize and utilize the body as an instrument of progressive political projects' (Hayes-Conroy and Hayes-Conroy 2010: 1277). Hence, political outcomes can eventuate from people learning how to sense in and across the practice of beekeeping.

This development to a more-than-human sensing can elicit wider consequences. Holmes Rolston III (1988) argues that the link between care and political action can influence citizens' conservation ethic, in which people who learn to care about nature are more eager to engage and sustain such activities. Learning through the senses can provide an important pathway to caring that, in turn, can contribute to engagement with, and conservation of, nature beyond beekeeping. The benefits of urban human and nature engagements are manifold, producing social, cultural, environmental, and economic outcomes (Frumkin 2003; Hinchliffe and Whatmore 2006; Maller *et al.* 2010). While the motivations and approaches for beekeeping differ – and indeed, where care is not always read as a 'positive' thing – by 'sensing in' to bees, I argue that 'a different kind of intelligence' (Hayes-Conroy and Hayes-Conroy 2010: 1275) can unsettle 'binary codings of human/nonhuman [that] dissolve into a coding of more-than-human' (Houston *et al.* 2017: 10). This dissolution provides steps towards achieving a more comprehensive understanding of the world (Maller 2018) where 'closer' human/nature engagements have transformative potential to foster new practices and knowledge (Tsing 2015: 20).

Sensing the multispecies city

Acknowledging the senses in urban beekeeping is multi-fold: it helps humans relate to nonhumans, it enables nonhumans to influence human behaviour and promote self-reflection, and through experiential learning, it can help humans relate to each 'other'. It also has very practical implications, raising concerns about the loss of senses in elderly beekeepers (who represent a majority of beekeepers), which can impact their ability to detect pests and disease in their hives. Furthermore, the senses offer insights for responding to the unique conditions produced by urban environments, such as light and noise pollution, a predominance of hard surfaces, and an assemblage of native and non-native species. Whilst this chapter has examined the domain of bees, I argue that the senses extend into, and should be examined in, the lifeworlds of other forms of urban livestock, in addition to animals (and insects) within the wider urban ecosystem that contribute to transforming cities into productive food sites.

Importantly, this sensual lens of knowledge enacted by beekeepers also enables 'alternative accounts of being or becoming in the world (ontology) as well as other means and methods of knowing and explaining it (epistemology)' (Maller 2018: 23). By 'bringing nature back' through the senses, beekeepers are ontologically broadening 'who' and 'what' has the right to the city, contributing to the imaginary of the 'multispecies city'. For a multispecies city to be realised, Iris Duhn (2017: 46) recognises that there will need to be 'an emerging emphasis on the ethics and politics of sharing spaces with others' in addition to 'a re-thinking of what place is, and who and what makes places (Duhn 2012; Steve Hinchliffe and Whatmore 2006; Tuck and McKenzie 2014; van Dooren and Rose 2012)' (ibid). A shift to a multispecies cities will also need to recognise the rights of both humans and nonhumans, opening new debates regarding access to space, the identification of one's needs, and active consideration for all urban residents in social and political decision-making processes (Houston *et al.* 2017; Shingne 2020). The senses add an essential element and approach in which to consider such a transition.

Conclusion

This chapter argued that a more-than-human methodology of the senses contributes to a holistic understanding of urban food production. The senses play a pivotal role in beekeeping, destabilising human and nonhuman boundaries, contributing greater understandings for creating convivial, multispecies cities. Hayes-Conroy and Hayes-Conroy's (2010) concept of visceral geography provided a useful frame to draw out the role of senses in beekeeping, in which 'mattering' is recognised through sensing the vibrant agency of bees in piquing peoples' participation, and 'relating' is apparent in both human body and mind where affect across species produces relational physical and psychological consequences impacting people's behaviour, reflection, and beyond-human understandings. 'Defying boundaries' is expressed through beekeepers' 'making sense' of bees to human 'others', with possible contributions to conservation efforts. By acknowledging the role and power of the senses, this chapter stressed their importance both in practice for new urban beekeepers and in theory for consideration in broader food and cities literature. Finally, I argued that the senses provide a pathway to reconsider conventional understandings of who and what cities are for and how cities can be reconceptualised beyond anthropocentric perspectives to acknowledge the rights and benefits of realising more convivial, multispecies worlds.

References

Atkins P. (2012) *Animal Cities: Beastly Urban Histories*. Burlington, VT: Ashgate.

Barth F. (2002) 'An Anthropology of Knowledge 1', *Current Anthropology* 43 (1): 1–18.

Bennett J. (2004) 'The Force of Things: Steps Toward an Ecology of Matter', *Political Theory* 32 (3), June: 347–372.

Bingham N. (2006) 'Bees, Butterflies, and Bacteria: Biotechnology and the Politics of Non-human Friendship', *Environment and Planning A* 38: 483–498.

Blecha J. (2007) 'Urban Life with Livestock: Performing Alternative Imaginaries Through Small-Scale Urban Livestock Agriculture in the United States', Unpublished Thesis, University of Minnesota, Minneapolis.

Brinkley C. and Vitiello D. (2014) 'From Farm to Nuisance: Animal Agriculture and the Rise of Planning Regulation', *Journal of Planning History*, May. DOI: 10.1177/1538513213507542.

Carrington D. (2019) 'Plummeting Insect Numbers "Threaten Collapse of Nature"', *Guardian*, 10 February, www.theguardian.com/environment/2019/feb/10/plummeting-insect-numbers-threaten-collapse-of-nature.

Cloke P. and Jones O. (2003) 'Grounding Ethical Mindfulness for/in Nature: Trees in Their Places', *Ethics, Place & Environment* 6 (3): 195–213.

Cockrall-King J. (2012) *Food and the City: Urban Agriculture and the New Food Revolution*. New York: Prometheus Books.

Duhn I. (2012) 'Places for Pedagogies, Pedagogies for Places', *Contemporary Issues in Early Childhood* 13 (2): 99–107.

Duhn I. (2017) 'Cosmopolitics of Place: Towards Urban Multispecies Living in Precarious Times', in Malone K., Truong S. and Gray T. (eds) *Reimagining Sustainability in Precarious Times*. Singapore: Springer: 45–57.

Edwards F. and Dixon J. (2016) 'The Hum of the Hive: Negotiating Conflict Between Humans and Honeybee Towards an Ecological City', *Society & Animals* 24 (6): 535–555.

Frumkin H. (2003) 'Healthy Places: Exploring the Evidence', *American Journal of Public Health* 93 (9): 1451–1456.

Goodman D., DuPuis M. and Goodman M. (2011) *Alternative Food Networks: Knowledge, Practice and Politics*. Abingdon and New York: Routledge.

Goodman M. (2015) 'Food Geographies I: Relational Foodscapes and the Busy-Ness of Being More-Than-Food', *Progress in Human Geography* 40 (2): 1–10.

Haraway D. J. (2008) *When Species Meet*. Minneapolis: University of Minnesota Press.

Hayes-Conroy J. and Hayes-Conroy A. (2010) Visceral Geographies, *Geography Compass* 4 (9): 1273–1283.

Hayes-Conroy A. and Hayes-Conroy J. (2010) 'Feeling Slow Food: Visceral Fieldwork and Symmetrical Research Relations in the Alternative Food Movement', *Geoforum* 41 (5): 734–742.

Hinchliffe S. and Whatmore S. (2006) 'Living Cities: Towards a Politics of Conviviality', *Science as Culture* 15 (2): 123–138.

Hitchings R. (2003) 'People, Plants and Performance: On Actor Network Theory and the Material Pleasures of the Private Garden', *Social & Cultural Geography* 4 (1): 99–114.

Houston D., Hillier J., MacCallum D., Steele W. and Byrne J. (2017) 'Make Kin, Not Cities! Multispecies Entanglements and "Becoming-World" in Planning Theory', *Planning Theory*. DOI: 10.1177/1473095216688042.

Hyvärinen P. (2019) 'Beekeeping Expertise as Situated Knowing in Precarious Multispecies Livelihoods', *Sosiologia* 56 (4): 365–381.

Ingold T. (2000) *The Perception of the Environment: Essays on Livelihood, Dwelling and Skill*. London: Routledge.

Lave J. (2011) *Apprenticeship in Critical Ethnographic Practice*. Chicago: University of Chicago Press.

Lave J. and Wenger E. (1991) *Situated Learning: Legitimate Peripheral Participation*. Cambridge: Cambridge University Press.

Maller C. (2018) *Healthy Urban Environments: More-Than-Human Theories*. Abingdon and New York: Routledge.

Maller C., Henderson-Wilson C. and Townsend M. (2010) 'Re-Discovering Nature in Everyday Settings: Or How to Create Healthy Environments and Healthy People', *Eco-Health* 6 (4): 553–556.

Mata L., Ramalho C. E., Kennedy J., Parris K. M., Valentine L., Miller M., Bekessy S., Hurley S. and Cumpston Z. (2019) 'Bringing Nature Back into Cities', *People and Nature* 2: 350–368.

McKie R. (2018) 'Where Have All Our Insects Gone?' *Guardian*, 17 June, www.theguardian.com/environment/2018/jun/17/where-have-insects-gone-climate-change-population-decline.

Moore L. J. and Kosut M. (2013) *Buzz: Urban Beekeeping and the Power of the Bee*. New York: New York University Press.

Pálsson G. (1994) 'Enskilment at Sea', *Man* 29, December: 901–927.

Philo C. and Wilbert C. (2000) *Animal Spaces, Beastly Places: New Geographies of Human-Animal Relations*. London: Routledge.

Puig de la Bellacasa M. (2017) *Matters of Care: Speculative Ethics in More Than Human Worlds*. Minneapolis: University of Minnesota Press.

Rolston H. III (1988) *Environmental Ethics: Duties to and Values in the Natural World*. Philadelphia: Temple University Press.

Shingne M. C. (2020) 'The More-Than-Human Right to the City: A Multispecies Reevaluation', *Journal of Urban Affairs*. DOI: 10.1080/07352166.2020.1734014.

Tuck E. and McKenzie M. (2014) *Place in Research: Theory, Methodologies, and Methods: Theory, Methodology, and Methods*. London: Routledge.

Tsing A. (2015) *The Mushroom at the End of the World: On the Possibility of Life in Capitalist Ruins*. Princeton, NJ: Princeton University Press.

van Dooren T. and Rose D. B. (2012) 'Storied-Places in a Multispecies City', *Humanimalia* 3 (2): 1–27.

Witmore C. L. (2004) 'Four Archaeological Engagements with Place: Mediating Bodily Experience Through Peripatetic Video', *Visual Anthropology Review* 20 (2): 57–72.

Wolch J. and Emel J. (eds) (1995) 'Theme Issue on "Bringing the Animals Back in"', *Environment and Planning D: Society and Space* 13: 631–760.

5 Sensing vernacular Chennai, not Madras – a photo-essay

Roos Gerritsen

Source: Photograph by Roos Gerritsen

1. Mint street food walkers

It is a Sunday, 14 February 2016, 9am. The temperature is still pleasant at this time of the day and year, and Sowcarpet is dissonantly quiet in comparison to other days. Mint Street is a narrow street in comparison to current roads, yet it is the widest street of Sowcarpet, an old neighbourhood in central Chennai: bikes, rickshaws, cows, and garbage make the street even narrower, particularly for pedestrians who need to navigate the hurdles on their path, giving way to more speedy road users. As a pedestrian, you take over the rhythm of the street, slowed down by its own rush of activity. Yet on this Sunday morning, Sowcarpet is waking up slowly. The velocity of movement is less: fewer bikes, fewer horns, barely any shops open, and one or two cycle rickshaw drivers asleep on their bikes. Gathered are a group of around ten foodies from a Facebook food walk group organised by Vishant. *Squelch.* Vishant tramps in

Source: Photograph by Roos Gerritsen

Source: Photograph by Roos Gerritsen

cow dung with his *chappal* (sandal), and the group needs to halt. Without spending words or making repulsive faces due to this unpleasant sensation on his foot, Vishant immediately takes a leaf from the garbage pile that we just passed and attempts to wipe the manure off his foot and shoe. Partly successful. So far, so good; his foot is still dirty but not totally covered anymore. We continue.

2. Patra

We continue to walk down Mint Street, heading towards Mansukhlal Mithaiwala (sweet shop) that sells *patra*, a celebrated Gujarati dish of steamed colocasia leaves that is at once sweet, spicy, and salty. *Patra* is only available in the mornings and will soon be sold out. Several of our group buy a portion, served on

Source: Photograph by Roos Gerritsen

two leaves and a piece of newspaper. *Click clack.* The *patra* is photographed from all sides with phones and SLR photo cameras. We discuss its taste and texture; its broad palette of flavours, from sweet to tangy; its softness and fluffiness; and the freshness of the coconut chutney. The latter is a recurring item in south Indian cuisine, and everyone knows how coconut chutney in restaurants can be eaten only when it is fresh. Some people take some home for their relatives; others warn that they have to eat it within a few hours, before the chutney goes bad.

3. Foodies

On a Saturday evening a few months later, Vishant and a group of the Facebook followers, venture out on another trip to Sowcarpet. *Patra* will not be available now. The evening, though, gives another sense-scape. As we wait at our meeting point at the busy junction of the Sri Renuka Parameswari Temple, the sun still shines briskly but will soon slide behind the three-to- five-storey buildings. The flower-seller sells jasmine and rose buds; worshippers buy the necessary items to offer to the goddess. *Ding ding ding ding.* The temple bells ring. Around the garbage pile that has been adding up behind a traffic police barrier at the other side of the temple, a small monkey, dressed up in a worn, torn costume and chained to the barrier, is repetitiously moving around with

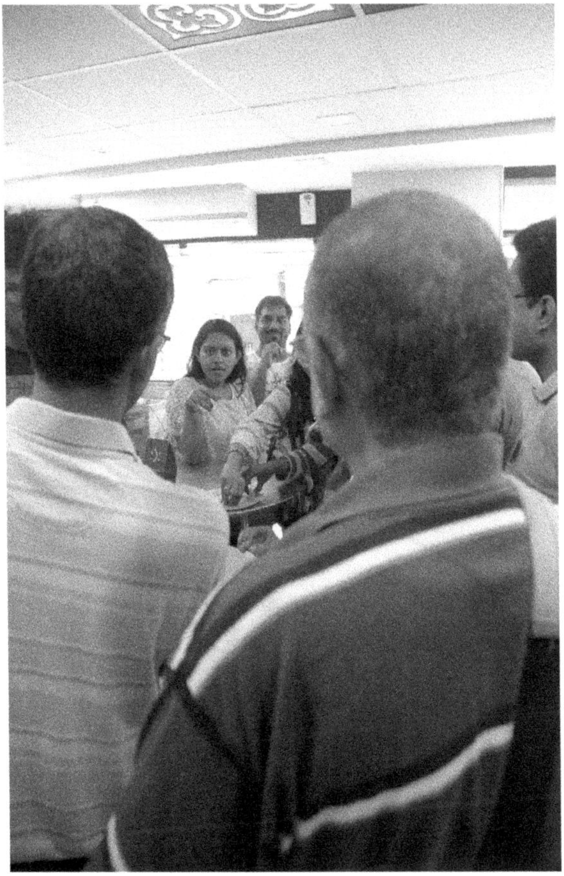

Source: Photograph by Roos Gerritsen

typical caged animal behaviour. A *keerai* (greens) seller pushes his cart through the street, calling out the various keerai that he has for sale. *Palak keerai! Arai keerai! Murungai keerai!* His voice evaporates within the loudness of traffic. A woman with a cloth shopping bag approaches the keerai seller to buy his greens, carefully looking at the different bundles, touching them to see which one is garden fresh.

There is barely a place to wait for all group members, as the number of bikes, cycle and auto rickshaws, and cars swells, dominating the soundscape with two-tact motors, horns, and impatient bells. A cow completely blocks traffic until a young man energetically makes an effort to move the cow away from the junction with pats on his back and loud appeals. The cow moves away, causing another stretch of the road a few metres away to be jammed. All members of the food walk have arrived now, dropped by an Ola cab, private car, or

Source: Photograph by Roos Gerritsen

motorbike a few hundred metres away. Sowcarpet itself cannot be entered by cars; the roads are too small and too jammed for cars to traverse them. Sweat is wiped off with *dupattas* (Indian shawl-like scarves) and handkerchiefs. We are ready to go.

4. The visual

The first photographs and text vignettes introduced in this chapter attend to food, city, and the senses beyond writing. While text accompanies the images, I encourage the reader to consider the images in their own right, to see them as more than mere illustrations of the text. This work is based on ten months of fieldwork on urban food practices in Chennai that focused on health, the city,

Source: Photograph by Roos Gerritsen

Source: Photograph by Roos Gerritsen

Source: Photograph by Roos Gerritsen

Source: Photograph by Roos Gerritsen

Source: Photograph by Roos Gerritsen

and sensorial explorations. For this photo-essay, I conducted participant observation during food walks, at street food markets, in restaurants, and at (mostly) middle-class homes. I interviewed a wide range of people, from restaurateurs and food shop owners to people who enjoy exploring new food experiences; I used photography and video during my fieldwork. Elsewhere, I have explored how several of the food practices emerging in Chennai are informed both by the pleasures and corporeal experiences of food and by a rising feeling of distrust, societal anxiety, and the expressed desire for healthier lives (Gerritsen 2020; see also Sutton 2010; Ray and Srinivas 2012). While I generally resort to writing as an accepted format for publication, visual methodologies and representation are crucial aspects of both fieldwork and output (Gerritsen 2019).

How do we explore, understand and convey the sensorial relationalities that food practices entail, and what other kinds of questions and knowledge does

Source: Photograph by Roos Gerritsen

a photo-essay bring us? Just as ethnographic research is exploratory, reflexive, and therefore partially open ended (Hammersley and Atkinson 2007), using photography as method and publication format must also respond to the unanticipated, qualitative nature of research. By exploring the limitations and possibilities of the photo-essay, I argue for a more inclusive multimodal methodology for grasping the nexus of food, city, and the senses.

Realising that in the visual we also encounter a limited framework 'to think with' the city and the senses (we can still not taste, smell, feel, or hear), I see the photo-essay as an additional format to experiment with. Experimenting with non-textual productions like photo-essays or artistic productions 'builds upon the disciplinary concerns about the politics and power of representation but is also influenced by the ideas and practices of science' (Cox *et al.* 2016: 3). Going

Source: Photograph by Roos Gerritsen

beyond text and through, for example, audio or visual technologies, Rupert Cox *et al.* argue, is not merely exploring experimental practice, methodology, or modes of representation but also being aware of the ways in which 'voice' might be politically and morally articulated. Similarly, I do not argue that the visual essay gives us a better or different understanding of a topic necessarily, but it allows for experimentation in the collection and representation of different kinds of knowledge, of different ways of experiencing the world. As Anna Grimshaw has argued, techniques of visual anthropology might 'render different kinds of ethnographic knowledge from the kinds that are articulated through the framework of a discursive anthropology' (Grimshaw 2005: 18). The observational cinema that Grimshaw sets out places one differently in the world and fosters a new awareness of the non-verbal, of movement, gesture, posture, action, refocusing attention around the details, textures, and materiality of the social world (op. cit. 23). In relation to this chapter, I pay specific attention not only to the sociality that informs the food walks, but also to the food walks as an activity that is both placed and sensed. Or, to use Steven Feld's words, 'as place is sensed, senses are placed; as places make sense, senses make place' (Feld 1996: 91). It is not only the way in which the visual creates awareness of the non-discursive; it also tries to capture the ways in which places and senses mutually engender each other. The purpose of the photo-essay here is to lay out another way of embodying these connections between place, senses, and food.

By visualising the food walks and the relation to the neighbourhood they are situated in, I attend to sensorial ways in which food walks are more than seeking taste. The photographs that I took have a double purpose. First of all, they gave me a way to 'record' and observe elements that would otherwise have gone unnoticed. They allowed me to pay attention to certain elements during the walk and, by systematically working with the pictures afterwards, to see different aspects later on. And second, the photographs allow me to think differently about the moments of interaction during such a tour. While participants had plenty of time to chat over and about the food items we ordered, about other interests in the social worlds of the participants, the chats were exceptionally fragmented due to the interruptive environment: crowds, noise, narrow streets, and the limited capacity of the small food stalls made conversations difficult to maintain. This disruptive experience, I argue, is more appropriately depicted in images, also while most participants communicated through their camera as well. Images were shared immediately or afterwards via the Facebook site or WhatsApp groups, depicting the group of foodies and the various food items we had seen, tasted, smelled, and touched. This, however, does not mean that the images tell an uninterrupted story. Instead, the images are meant to display the interruptive experience of eating, talking, and walking as well. The essay is an attempt to convey experiences beyond the mental facility of language (MacDougall 2005: 2).

The private Facebook group set up by Vishant counts 32,032 members.[1] Vishant, working in IT consultancy and an active Chennai citizen, started the Facebook group a few years ago with a friend, out of their interest in exploring different eateries in the city. He does it out of his personal interest, and anyone who wants to join is welcome. He usually announces a day and time for the walk or asks members of the group where they want to go next time. For food walks, usually between 10 to 50 people participate. They come from all parts of town, mostly from more affluent neighbourhoods, aiming to explore a neighbourhood and its food that they do not otherwise visit that often, or at least not for that purpose. Most attendees are regulars, living in Chennai. Participants often come alone or sometimes with their family. One of them is Sanjay, a corpulent man in his 40s who is always on the lookout for good food. When I meet him later under a menu of a *murukkku*[2] sandwich, a current street food hit in Chennai, he explains how the food tours for him are about meeting people as well:

> It's nice you know, to eat food together, and explore neighbourhoods. I know these places in the meantime, but the tours are a way of finding out about places you would otherwise not travel to for food or social meetups.

Other participants also signalled this interest in exploring city neighbourhoods. I met an IT expert who worked almost two hours travelling away to the southern part of town and who realised one day that he never had time to actually *be* in the city. He travels every day through town for work but never stops to eat, observe, or explore. The food walks, he explained, were a way to change that. The slow pace of walking, talking to others, and getting to know well-known food joints in neighbourhoods were a way, he suggested, to explore and get to know the city. They helped slow down his working life. Many of the participants come back for walks, alone or with a friend or family members, in neighbourhoods that they have explored previously. I will describe next how such explorations are a way to explore typical food and the place this food is situated in. Without the comfortable routine of other moments of conviviality in restaurants or eateries, for example, going by rickshaw or car and being dropped off in front of the door, keeping walking to a minimum, these walks are about really being in that place and not only about being projected into it. In this way, food is sensed as much as the environment is sensed.

For a long time, food in India has been predominantly described in what Amita Baviskar has aptly called the conventional containers of caste-village-family (2019: 368). Recently, more scholars have been working on themes around food in India which exceed these conventional containers and have presented a more nuanced sociality of city life, movement between places, and social relationships (Appadurai 1981; Caplan 2008; Khare and Rao 1986; Ray and Srinivas 2012; Solomon 2016; Staples 2014). I follow this body of work and set aside the conventional approaches of understanding sociality in India, looking specifically at the place that brings different people and imaginations together.

I will first go back in time to situate the neighbourhood Sowcarpet, where the food walks that I started the chapter with took place. Then I will suggest how this idea of the vernacular city, the non-elite or the common, which is normally not observed as the image of the city (Venkatachalapathy 2006), creates an image of the 'real' city, of an experience that goes beyond a regular middle-class lifestyle (Brosius 2010; Caplan 2008; Donner 2008; Fernandes 2006).

5. Sowcarpet

Sowcarpet has a specific image in the imagination of Chennai citizens. Sowcarpet is a neighbourhood in northern Chennai that is historically

Source: Photograph by Roos Gerritsen

part of the 'black town' of colonial Madras. Madras, the first colonial city in India, was founded in 1639 when Francis Day, working for the East India Company, bought a piece of land. Madras consisted of a white part where the East India personnel settled, first in Fort St George, and the so-called Black Town, where non-Europeans lived. Let me quote a historic description from S. Muthiah, a well-known Madras/Chennai historian and journalist:

> Thomas Salmon, writing in 1699, stated that 'Black Town', "where the Portuguese, Indians, Armenians, and a great variety of other people inhabit . . . is built in the form of a square . . . better than a mile and a half in circumference; being surrounded with a brick wall seventeen feet thick . . . The streets of the Black Town are wide, and trees are planted in some of them; and having the sea on one side and a river on the other, there are few towns so pleasantly situated or better supplied; but except some few brick houses the rest are miserable cottages, built with clay and thatched and not so much as a window to be seen on the outside . . . but I must say, notwithstanding all this appearance of poverty, I never was in a place where wealth abounded more, or where ready money was more plentiful about twenty years ago . . . Beyond the Black Town are gardens for half a mile together planted with mangoes, coconuts, guavas,

Source: Photograph by Roos Gerritsen

> oranges . . . where everybody has the liberty of walking and may purchase the most delicious fruits for a trifle.
>
> (Muthiah 2004: 325)

While the colonial city developed as Madras, the name Chennai, or Chennapatnam as the native town of colonial Madraspatnam was called, existed parallel to Madras as its mirror image. The city changed its name to Chennai in 1996, in line with other 'nativist' agendas that claimed the desire to vernacularise the city and to decolonise it (Arabindoo 2006): Bombay became Mumbai, Calcutta became Kolkata, and Bangalore became Bengaluru. In Chennai, the words *Madras* and *Chennai* also mark a distinction of inhabitants, in which Chennai is the vernacular city, the non-elite city that is often left out in its cultural history and that is not imagined as what the city is (Venkatachalapathy

2006). Sowcarpet has grown out of the Black Town, a part of the city that is central to its formation but often left out in its imagination.

Sowcarpet developed into a commercial area hosting most of its wholesale trade. The neighbourhood is divided in sections where retailers sell hardware, cooking utensils, religious paraphernalia like Kumkum powder, and brass lamps. It is a congested area, and most Chennai residents only visit it if they need to buy from the retailers. For many Chennaiites in other parts of the city, Sowcarpet is the manifestation of the vernacular city. Another quote from S. Mutiah:

> Between General Hospital in the south and the remains of the great wall in the north lies the George Town of today – busy public institutions in the south and east, a hive of trading activity in the centre and a crowded residential quarter in the north, where the roadside homes are today much as they were in 1699 when Ensign Thomas Salmon was stationed here and recorded, on his return to England. . . 'brick houses . . . of the better sort . . . are of the same materials and built usually in one Form, that is with a little square in the middle from which they receive all their light'. . . . And in these homes, the 'Gentus' and Malabarrs – the Telugus and Tamils of George Town today – live life very much as it must have been in 1674, when there were only 75 houses in 'Black Town' or in 1750 when there were 8,700. In 1855, a Gazetteer recorded, 'The minor streets . . . are numerous . . . extremely narrow . . . the form of the house (here) resembles . . . a hollow square, the rooms opening into a courtyard in the centre, which is entered by one door, from the street. This effectually secures the privacy so much desiderated by the Natives, but at the same time prevents proper ventilation, and is the source of many diseases. The street, with few exceptions, have drains on both side which are deep and narrow'. . . . Little changes in George Town, it just keeps getting more and more congested. . . . [T]he southern half of the 'city within', mainly in Peddanaickenpet, is Madras's wholesale market, street after street, narrow and crowded, specializing on street level in particular goods, with palatially-equipped homes often occupying the upper reaches of the same dingy buildings.
>
> (Muthiah 2004: 323–324)

The description of the unchanged, congested neighbourhood of the 'Black Town' is not only an interesting view back in time; it also seems to confirm the idea of the vernacular city, then and now. The idea of an unchanged neighbourhood raises the impression of a place one needs to discover as a glance back in time, a glance giving insight into the vernacular town.

6. Chennai, not Madras

Back to the present. Due to the long-term presence of Rajasthani and Gujarathi trader communities, Sowcarpet hosts many restaurants that offer food

Source: Photograph by Roos Gerritsen

for and from such communities. While most middle-class neighbourhoods are generally more stretched out and set up, Sowcarpet also hosts economically well-off inhabitants, but this doesn't necessarily show in the urban fabric, condition of the buildings, or neighbourhood infrastructure.

The topographical and ethnic stratification can be mapped on the vernacular topography of the city (Manalansan 2006). Sowcarpet, and North Chennai at large, is more congested due to narrower streets and has less infrastructure than the southern part of the city (Arabindoo 2006; Beelen *et al.* 2010). As Pushpa Arabindoo has observed, it comes with a certain irony that the state politics of vernacularising Chennai come with an uneven geographic development, in which south Chennai 'exhibits an elegant and ordered landscape . . . replete with face-lift, while the northern half portrays filth and decay, and poor infrastructure' (2006: 31). North Chennai translates into another kind of sensory

Source: Photograph by Roos Gerritsen

scape created by narrow lanes, congestion, loudness, dust, and no proper garbage collection, to name just a few.

Sowcarpet is envisaged as the vernacular city, as the origin of Chennai, against the image of colonial Madras. Nonetheless, it is hardly present as more than an image for many. It is mostly Mylapore, another neighbourhood whose history dates back to the pre-British and Portuguese presence, that people religiously and culturally relate to. Mylapore accommodates the well-known Kapaleeshwarar temple devoted to Shiva, and the historic presence of a large Brahmin community makes it an important part of the religious ideoscape. Georgetown, on the other hand, while seen as the inception of Chennai, is the place that brings economic relations but not necessarily cultural ones. One visits Sowcarpet to buy something specific, but one hardly ever visits on cultural, social, or religious grounds.

Source: Photograph by Roos Gerritsen

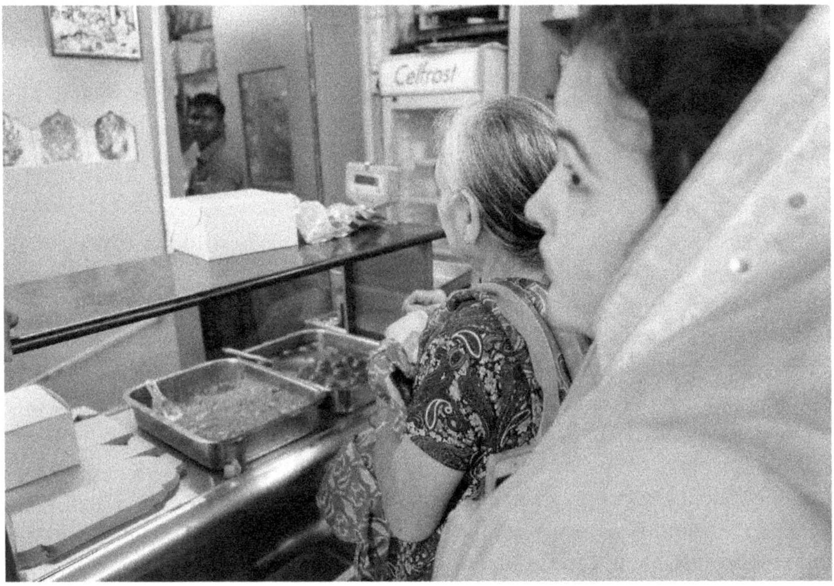

Source: Photograph by Roos Gerritsen

Source: Photograph by Roos Gerritsen

During the food walks, there is an interest in the other side of town due to its otherness. In contrast to the 'smelly immigrants' that Martin Manalansan (2006) describes, the inhabitants of George Town are actually made synonymous with the food that is sold, something that is tasty, full of aroma, and visually attractive. The food walks, I argue, have become a means to actually find a cultural common ground by means of food. Experiencing the neighbourhood while walking around, checking out the rose milk, *jalebis*, *fafda*, *patra*, and special idli are ways to get to know the vernacular city. While many of the food items available are familiar items on the menu in south India, the strong presence of communities from Rajasthan, Andhra Pradesh, and Gujarat also makes the range of available food much wider. From religious dietary restrictions to specific regional preparations or items, the food in Sowcarpet is largely distinctive to what most of the food walk attendees eat at home. Hence, instead of the mnemonic dimensions of sensory experiences of migrants (Sutton 2001; Law 2001), we can discern an opposite enframement of people actively seeking the sensory memory of the city, attending to a closeness of that unknown part of the city through flânerie, sociality, and food consumption. It is therefore not only a certain idea of history that is revived, but also the vernacular city. Locality, Harris Solomon has argued, is made, not given, and similarly, the streets in which the consumption of food and sociality take place are not only backdrops or empty containers but actively part of the experience of the city (2015). Similarly, food is one of the infrastructures that make up the city.

7. *Thattu idli* at Jai Sri Vaishnavi's

Back to one of the walks. One of us ordered *thattu idli*, a steamed dumpling made of a fermented lentil-rice batter. The cook in this small shop in a back street of Mint Street drizzles a generous amount of ghee over the idli and hands over the plate. The stall is well known for its *thattu idli* ('plate idli', a larger version of the ordinary idli) and has an unusual selection of exotic-sounding *dosas* (the pancake variant made of the same batter), like the Mexican, Italian, and chocolate *dosa*. I ask the vendor what a Mexican or Italian *dosa* consists of. It turns out they are all more or less the same *dosa*s, covered with cheese. We decide to order the thattu idli instead, doused in ghee and served with sambar (a lentil-based stew that is a basic in Tamil Nadu cuisine), *podi* (a powder of spices, lentils, and peanuts), and chutney. We all take a bite from one plate,

Source: Photograph by Roos Gerritsen

using our right hand. As our hands get oily from the ghee-idli, I struggle to use my camera to take pictures. Others are clearly more experienced and take their picture first before setting to their idli. It's not only much easier, but also makes much better pictures of the whole dish than the half-eaten, spilled-sambar one that I photographed. This street is residential and less loud, but the group has to eat their *idli* or *dosa* on different ends of the street due to the many motorbikes parked on the road. Such streets clearly mark the urban challenge of motorised vehicles, which is currently under even more pressure due to the exponential increase in car ownership. We now have some time and space to chat a bit, and we hang around a bit longer than we do at the other locations at which we stop. No one really knows Sowcarpet well, but several participants of the tour are now familiar with various locations where one gets specific items. The owners of these food outlets have noticed that there are more visitors from 'outside'; some have TripAdvisor stickers or newspaper articles in which they appeared on their windows as markers of their special position. The food walks seek a sense of place, but they have also created place.

The urban environment is not a backdrop against which certain practices take place. It attracts bodies and things; it is part of the way in which subjectivities are formed and formulated. The sounds reverberating in the narrow streets of Sowcarpet are loud – too loud, according to a study by the Central Pollution Control Board that mentioned that the noise levels recorded in Sowcarpet and various other city zones were double the permissible limits (Raghunathan 2017). Cycle rickshaws ringing their bells, two-tact engines of auto rickshaws, motorbikes. They mostly carry products, and some carry people. The economically vibrant neighbourhood is in motion throughout the day. Small vans come and go to deliver or pick up goods. The sounds around food are less easy to sense. For these sounds – the sizzling of oil, the stainless-steel utensils, the cutting of onions, the itinerant vendor riding his cart through the street – one needs to block out traffic sounds. The familiar hisses of pressure cookers cooking rice in homes are barely to be heard here.

8. Synesthetic environments, synesthetic experiences

While the photographs themselves do not necessarily give us clear 'representation' of how food walks are sensorially informed, they do give us a narrative that is not merely discursive. In this chapter, I have attempted to show that it is not the images per se that give us methodological clues about the ways in which we need to 'capture' food in the city through the senses. Instead, I hope to have shown that senses are placed, and places are sensed in ways that are diverse and changing according to time, place, and subjectivity. The foodies, as they explore Sowcarpet, an old neighbourhood in Chennai, are actively seeking new experiences that are led by the senses. As a group focused on food, they not only sense food; they sense the neighbourhood as well. And the latter experience is crucial to sensing the food. However, this is not necessarily a quest for taste alone, but a synesthetic experience, in which a place is made

Source: Photograph by Roos Gerritsen

sense of through the multi-sensorial experience that includes the local conditions of sensation, knowledge, and imagination (Feld 1996: 91).

As Ferne Edwards states in this volume (Chapter 4), a methodology of the senses asks to stop for a moment to simply listen. Here, I could ask a similar question on the level of the photo-essay itself and see what simply watching can bring us. Rather than seeing the different modalities of a place as 'noise', we can see them as articulations that are sensed and make sense in relation to the walks. The food is sensed through these modalities. Looking at these experiences of a multi-sensuous food walk, in which taste, space and, sociality come together, we can understand how the senses cannot be seen as separate, individual sensualities but should be seen in relation to the sociality in which they occur. Edward Casey argues that the structures of place that permeate places are the experiencing bodies to place and second the ability of places to draw together bodies, things, time, and space (Casey 1996: 44). In this way, places are envisaged as continually changing. Following Edward Casey, Sarah Pink has suggested that such an approach allows us to consider place-making processes as parallel, interrelated, and overlapping place-making (Pink 2008). And this is exactly what the food walks do. On the one hand, they create the places they are in, yet they are also exploring a place, trying to experience it as it is. This overlapping and parallel place-making process is based on the imagination, knowledge, and sense of that place. And in this way, Sowcarpet, as the icon of vernacular Chennai, of an unfamiliar Chennai, becomes a little bit more made sense of.

Notes

1 Viewed in November 2019.
2 A sandwich of vegetables and chutney held by two small *murukku* (a rice flour and *urad dhal* deep-fried savoury snack, normally not used for making sandwiches but as a snack on its own).

References

Appadurai A. (1981) 'Gastro-Politics in Hindu South Asia', *American Ethnologist* 8 (3): 494–511.

Arabindoo P. (2006) 'Geography of a Lingua Franca: History of a Linguistic Fracas', in Venkatachalapathy A. R. (ed) *Chennai Not Madras Perspectives on the City*. Mumbai: Marg Publications: 19–38.

Baviskar A. (2019) 'New Cultures of Food Studies', in Srivastava S., Arif Y. and Abraham J. (eds) *Critical Themes in Indian Sociology*. New Delhi: Sage: 361–374.

Beelen K., Gerritsen R. and Srivathsan A. (2010) *Climbing Up the Rank Ladder*. Monu: Magazine on Urbanism.

Brosius C. (2010) *India's Middle Class: New Forms of Urban Leisure, Consumption and Prosperity*. New Delhi: Routledge.

Caplan P. (2008) 'Crossing the Veg/Non-Veg Divide: Commensality and Sociality Among the Middle Classes in Madras/Chennai', *South Asia: Journal of South Asian Studies* 31 (1): 118–142.

Casey E. (1996) 'How to Get from Space to Place in a Fairly Short Stretch of Time', in Feld S. and Basso K. H. (eds) *Senses of Place*. Santa Fe, NM: School of American Research Press.

Cox R., Irving A. and Wright C. (2016) 'Introduction: The Sense of the Senses', in Cox R., Irving A. and Wright C. (eds) *Beyond Text? Critical Practices and Sensory Anthropology*. Manchester: Manchester University Press: 1–19.

Donner H. (2008) 'New Vegetarianism: Food, Gender and Neo-Liberal Regimes in Bengali Middle-Class Families', *South Asia: Journal of South Asian Studies* 31 (1): 143–169.

Feld S. (1996) 'Waterfalls of Song: An Acoustemology of Pace Resounding in Bosavi, Papua New Guinea', in Basso K. and Feld S. (eds) *Senses of Place*. Santa Fe, NM: School of American Research Press.

Fernandes L. (2006) *India's New Middle Class: Democratic Politics in an Era of Economic Reform*. Minneapolis: University of Minnesota Press.

Gerritsen R. (2019) *Intimate Visualities and the Politics of Fandom*. Amsterdam: Amsterdam University Press.

Gerritsen R. (2020) 'Food Walks and Street Doctors: Health and Culinary Nostalgia in a South Indian City', in Crenn C., Matta R. and de Suremain C. (eds) *Food Identities at Home and on the Move: Explorations at the Intersection of Food, Belonging and Dwelling*. London: Routledge.

Grimshaw A. (2005) 'Eyeing the Field: New Horizons for Visual Anthropology', in Grimshaw A. and Ravetz A. (eds) *Visualizing Anthropology: Experimenting with Image-Based Ethnography*. Portland: Intellect: 17–30.

Hammersley M. and Atkinson P. (2007) *Ethnography: Principles in Practice*. London: Tavistock.

Khare R. S. and Rao M. S. A. (eds) (1986) *Food, Society, and Culture: Aspects in South Asian Food Systems*. Durham, NC: Carolina Academic Press.

Law L. (2001) 'Home Cooking: Filipino Women and Geographies of the Senses in Hong Kong', *Cultural Geographies* 8 (3): 264–283.

MacDougall D. (2005) *The Corporeal Image: Film, Ethnography, and the Senses*. Princeton, NJ: Princeton University Press.

Manalansan M. F. (2006) 'Immigrant Lives and the Politics of Olfaction in the Global City', in Drobnick J. (ed) *The Smell Culture Reader*. Oxford and New York: Berg: 41–52.

Muthiah S. (2004) *Madras Rediscovered*. Chennai: East West Books (Madras) Pvt. Ltd.

Pink S. (2008) 'Sense and Sustainability: The Case of the Slow City Movement', *Local Environment* 13 (2): 95–106.

Raghunathan J. (2017) 'Chennai Newsreel', www.madrasmusings.com/vol-26-no-20/chennai-newsreel/.

Ray K. and Srinivas T. (2012) *Curried Cultures: Globalization, Food, and South Asia*. Oakland: University of California Press.

Solomon H. (2015) '"The Taste No Chef Can Give": Processing Street Food in Mumbai', *Cultural Anthropology* 30 (1): 65–90.

Solomon H. (2016) *Metabolic Living: Food, Fat and the Absorption of Illness in India*. Durham and London: Duke University Press.

Staples J. (2014) 'Civilising Tastes: From Caste to Class in South Indian Foodways', in Klein J. and Anne Murcott A. (eds) *Food Consumption in Global Perspective: Essays in the Anthropology of Food in Honour of Jack Goody*. New York: Palgrave Macmillan.

Sutton D. E. (2001) *Remembrance of Repasts: An Anthropology of Food and Memory*. Oxford and New York: Berg.

Sutton D. E. (2010) 'Food and the Senses', *Annual Review of Anthropology* 39: 209–223.

Venkatachalapathy A. R. (2006) *Chennai Not Madras Perspectives on the City*. Mumbai: Marg Publications.

Part II

Past in the present

Memory and food

6 The sensorial life of *amba*

Taste, smell, and culinary nostalgia for Iraqi Jews in London and Israel

Joel R. Hart and Daniel Monterescu

Amba foodways: a pungent sensory history

A seemingly innocent, tangy condiment – one popular in Iraqi, Israeli, Palestinian, and Indian cuisines – tells the story of how ethnicity, class, and gender cross and are reconstituted in the Middle East and beyond. Described as 'the Iraqi version of mango chutney' (Berg 2019: 77), *amba* is hard to ignore. Its potent smell, which stems from the addition of fenugreek, turmeric, and vinegar to mango (*amba* means mango in the Indian language of Marathi), is retained in perspiration and stirs controversy among men and women, Ashkenazi and Mizrahi Jews, Israelis and Palestinians, Iraqis in Iraq, and in the diaspora. Its illusive quality persists among the diverse consumers and producers we encountered, who are reluctant to endorse it fully as a national culinary icon. It is Iraqi yet celebrated mainly by the diaspora; Jewish, but not quite; of Indian and colonial origins but markedly different; Israeli but from elsewhere; consumed by some Palestinians as popular street food due to the Israeli culinary exchange between Arabs and Jews in Israel/Palestine. *Amba* straddles culinary registers and challenges the very notion of place-based authenticity. At the same time, due to its powerful sensorial imprint – described by Israeli chef Meir Adoni as 'the umami of the Middle East' – it is both the basis of a culinary nostalgia for Iraqi Jews in Israel and London and a clandestine obsession for a peculiar community of taste (Swislocki 2008). As Anna Tsing writes, 'smell draws us into the entangled threads of memory and possibility' (Tsing 2015: 47).

The origins of *amba* reflect tortuous foodways across the Indian Ocean. According to a common urban legend, members of the Jewish Baghdadi Sassoon family of Bombay invented *amba* in the mid-nineteenth century. Their first encounter with the luscious mango led them to send barrels of it coated in vinegar to Basra port, thus confirming its role in the story of the Jewish culinary diaspora, with roots in Iraq. *Amba* was part of a broader repertoire of pickles (*turshi*) consumed by Iraqis, yet in contrast to other forms of *turshi*, *amba* was largely consumed and purchased on the streets, provided by vendors and spice stalls at the Shorja Market. The reason for this is that the pre-pickled mango was imported from India and came in barrels to Basra and onwards to Baghdad and other Iraqi cities. By the 1950s, the Indian company The Ship exported it in glass jars to Iraq, including a superior brand of complete small green mangoes in brine that one could purchase from specific vendors in the market.

Amongst Iraqi Jews, *amba* had a special prominence, ritualised into Shabbat culinary traditions. *Amba* was typically eaten with eggs that slow-cooked in the same pot as chicken and rice in the most classic Iraqi-Jewish Shabbat dish, the *t'beet*. For breakfast, after the morning service, families ate sliced eggs on *khubz* (bread), with fried aubergine, parsley, and *amba*. *Batinjan wa beid tbeet* is a Jewish culinary recipe that would later reincarnate into the *sabih*, a famous street food in Israel, which adds tahini and salad to fried aubergine and *amba* and stuffs it all in a pita bread. With the rising popularity of the *sabih* sandwich in the 1980s, *amba* transcended the Iraqi domestic sphere and became a popular street food. Iraqis increasingly used *amba* as a condiment for shawarma and falafel, balancing their fattiness with its savoury-sourness. This process, which we call the 'falafelisation of the *sabih*', established the status of *amba* as a key component of street food across cities in Israel.

Today, in urban Israel and many North American and European metropolises, thousands of restaurants and fast food stands serve *amba* along with tahini and pickles; some people even replace ketchup with *amba*, using it as a dip for French fries (Figure 6.1). It features in an eclectic and hybrid appropriation of migrant culinary heritage. As Ari Ariel (2019: 107) notes: 'A typical Israeli lunch might include Viennese schnitzel happily coexisting with a side dish of couscous and spiced with a dash of Iraqi *amba* or Yemenite *zehug* (hot sauce)'. For the Jewish Iraqi diaspora based in London, *amba* talk takes one down memory lane. *Amba* invokes aromas of bygone days in Baghdad, where *amba* eaten on a slice of *sammoun* diamond-shaped bread in school was associated with childhood (Somekh 2007). Out of context and torn from the quotidian aromascape, *amba* was re-domesticated in the diasporic imagination as a symbol of the past, consumed in the present only within the confines of the family and particular culinary establishments. Outside the Iraqi community, however, *amba* is part of the gourmetisation of contemporary Israeli and Middle Eastern cuisine within global cities.

This chapter follows *amba* to chart its history of migration and reception in cities, focusing on its role as a sensorial mediator of memory and belonging. Drawing insight from Ian Cook *et al.*'s (2004) 'Follow the Thing: Papaya', we conducted multi-sited research into the transnational geography of *amba*, including its politics, sociality, and sensoriality of consumption. While Cook focuses on the 'fetishism of the market', we trace *amba*'s migratory paths to demonstrate how people inscribe meanings through its forms, uses, and trajectories. We draw on Arjun Appadurai's theory of value and track the 'social lives of commodities' (1986) by tracing the flows of commodities as they pass through different 'regimes of value in space and time' (1986: 4). From this diasporic outlook, the symbolic and material value of *amba* refers not to actual consumption in Baghdad but to the memory of flavours and aromas. Beyond that, there are a significant number of non-Iraqi consumers who now make the

Figure 6.1 Amba in lieu of ketchup on French fries

Source: Photograph by Daniel Monterescu

lion's share of *amba* aficionados. Fieldwork was conducted from 2017 to 2018 and consisted of visits to restaurants and producers; interviews with importers, suppliers, and customers; and conversations with chefs and food scholars in three main sites: Israel (Ramat-Gan, Jaffa, Tel-Aviv, Or Yehuda, and Hatikva Market), the West Bank (Beit-Jala), and London. In addition, we surveyed online food communities and other textual and visual sources such as memoirs, cookbooks, and films. We begin with the narratives of Iraqi Jews in London, for whom *amba* embodies a visceral culinary nostalgia connecting them to a lost homeland. We then move to Israel, where we observe *amba*'s diffusion through distinctive registers of nostalgia, adoration, and recreation, culminating in its cult status today amongst many (mostly male) Israelis.

Diasporic practices of culinary nostalgia

For many Iraqis, *amba*'s taste and smell evoke memories of a lost homeland and an opportunity to dwell on them. Drawing on Swislocki's concept of culinary nostalgia, defined as 'the recollection or purposive evocation of another time and place through food' (2008: 1), we explore how *amba* takes on a sensory form of memory that connects body to place in the diaspora. After an agreement between the Israeli and Iraqi government in 1951, the overwhelming majority of the historic Iraqi-Jewish community, which numbered 135,000, relocated to Israel (Bashkin 2012). This abrupt turn of events was the product of a confluence of factors that included increased Arab nationalist activity after the 1948 Israeli-Arab War, anti-Communist measures by the Iraqi government targeting Jewish communists, and Zionist underground activity in Baghdad. Less well documented was the distinctive community who stayed in Baghdad, determined to continue building a multicultural, inter-communal modern nation-state and to maintain a Jewish presence in a city whose rhythm had been dictated by Jewish customs. Most Iraqi Jews who came to London in the 1960s and 1970s were part of the 6,000 who had stayed behind. They had continued to live prosperous, cosmopolitan lifestyles until 1963, when the Baʿath party came to power, and a pan-Arab vision superseded inclusive cross-communal versions of Iraqi nationalism. Discrimination and repression developed rapidly in Iraq and reached unprecedented levels in 1969 with the hangings of nine Jewish men. This event led to almost the entire community leaving by the mid-1970s to London, New York, Montreal, and Los Angeles. For many, London nurtured a new exilic culture of Iraqi Jews determined to maintain their distinctive Babylonian identity distinguishable from their Sephardic counterparts (Hart 2016).

Our focus here, however, is on the specific mnemonic processes by which diasporic connections are made through culinary practice, which positions *amba* consumption as a mnemonic device. The most common way for immigrants to maintain a connection to a lost city is through the palate, because:

> The encounter with even the faintest of flavours from elsewhere or another time – a peach, or a sprig of basil left in a glass on the windowsill – enable migrants to rekindle cultural memories, complete with a strong sense of emotional attachment, in the present. In this sense, the nose and taste buds help anchor dislocated bodies, by locating moorings amidst the most vaporous of materials.
>
> (Rhys-Taylor 2014: 8)

Yet for Iraqi Jews, whose exile is often accompanied by trauma, culinary nostalgia evokes painful memories. Whilst a diasporic Babylonian identity developed in London, Linda Dangoor notes in her Iraqi cookbook *Flavours of Babylon* (2011) her unease in adapting to this new community: 'When I came here, I wanted to forget my past. I was unEnglish. So I didn't want anything to do with Iraqi tradition for a long time'. However, her cookbook begins with the words 'For those of us who have been uprooted and exiled from our homeland, food, like language and music, becomes that familiar space in which we can recreate a sense of belonging and a sense of community' (2011: 2). Coming to terms with her diasporic past, Dangoor gradually reconnected with her roots. She told us:

> I found that I made peace with my past through food. It's an amazing thing. And it has to do with identity. I felt that once you leave your country, and you cannot use your language, the Arabic that I loved. And then you lose the weather, you lose the way of life, the thing that remains really is food.

It is through culinary culture, we argue, that Dangoor sensorially retained a visceral sense of place and belonging, even when conscious reflections on identity sought to forget her Iraqi past. Her persistent culinary nostalgia denotes a sensory form of memory that connects the body to place in the diaspora. Jewish-Iraqi scholar Sasson Somekh illustrates in his memories from Baghdad how the materiality of memory persists: 'the intoxicating taste of *amba* and the fresh bread was anchored under my tongue' (2007: 87). Despite a melancholic reality of rupture in the narrative of many Iraqi-Jewish exiles, the recalling of eating *sammoun wa-amba* from the vendors of al-Rasheed Street nostalgically connects them to their lost city and a sense of comfort in defying parental orders to observe, smell, and taste the street.

Nostalgia derives from the Greek *nostos* (to return home) and *algia* (pain), posing a 'dual archaeology of memory and place' (Boym 2001: xviii). In her consideration of memory in the imagination of place, Svetlana Boym delineates two types of nostalgias (ibid). Restorative nostalgia stresses the *nostos* and attempts a transhistorical reconstitution of a lost home, while reflective nostalgia thrives in the *algia*, the painful longing itself, and delays the homecoming, 'wistfully, ironically, desperately' (Boym 2001: xviii). For Iraqi Jews, the impossibility of return and the fate of distance makes the form of nostalgia almost exclusively reflective. But this occurs within 'memory work, a dialectical notion . . . that makes it possible to think remembering and forgetting together' (Fabian 2007: 78). As exhibited by Dangoor, *amba* (and Iraqi food in general) maintains ambiguous connections, even in the face of conscious efforts to forget many aspects of life in Iraq. Such connections are visceral, denoting embodied, unspoken experiences of nostalgia rather than continuous and conscious engagements with the imagined past. Akin to Boym's reflective nostalgia, we argue that many Iraqi Jews in London are distinctively defined by sensory connections that 'do not follow a single plot' (Boym 2001: xviii). Connecting to a lost place through the joys of consumption whilst actively trying to disconnect from the political trauma of that lost place is to dwell in 'ambivalence' (ibid).

Emile Cohen, a cultural activist seeking to preserve traditional Iraqi *Maqam* music, exhibits a complex relationship with his memory of Baghdad. Emile fondly shared with us his intricate memories of the 'so modern' house that his father built in 1954 and its neighbourhood of Karada; of eating *qaima* for breakfast, a thick curd made from buffalo milk; and the art club where he watched popular music concerts from his roof. In our conversation, he told us:

> The whole *amba* comes in barrels. In schools, it was quite common that we'd have *sammoun* and *amba*. You know, it was very cheap. So we were at school, and we were all like 'ohh', and Abu Karem takes a ladle of *amba* into the *sammoun*. Bang. And you eat that. That's what we would eat all the time.

Despite his vivid memories, Emile does not wish to return to Baghdad: 'I would rather live with the imagination, with the memories that I had'. Return would cast him as an 'exile amongst ghosts' (Loizos 1999: 258). His reflective nostalgia 'loves details, not symbols' (Boym 2001: xviii). He thus exemplifies what Boym calls 'diasporic intimacy': a knowledge that 'objects and places were lost in the past and . . . that they can be lost again' (2001: 255). For those experiencing such a deep uprooting, 'the illusion of complete belonging has been shattered' (ibid). Despite memories being embodied in a material imagination of a particular urban landscape – a place – the object of longing for home is not necessarily for this landscape, but rather for a sense of 'intimacy with the world' (Boym 2001: 251). Diasporic intimacy emerges from Iraqi Jews' awareness of a distinctly cosmopolitan heritage. Narratives are often interwoven with an appreciation of Iraq's uniquely Iraqi non-sectarian nationalism incorporating religious and ethnic diversity (Bashkin 2012: 7) and the particular inter-communal lifestyles of Iraqi Jews and Muslims before 1948 or the 1963 Revolution.

Amba thus becomes symbolic of this distinctly cosmopolitan heritage, viscerally connecting Cohen, Dangoor, and others to a sense of place and cultural legacy. Yet simultaneously, this continuity of *amba* consumption reproduces a sense of detachment, reinforcing the need for reflective nostalgia. There is a melancholic rupture in their narrative, and such culinary nostalgia does not distinctly provide 'frameworks for articulating both ideology and utopia' (Swislocki 2008: 5). Memories are thick and visceral, but the practice of *amba* in London today is tied to the reality of displacement. Refraining from physically visiting Baghdad, the Jewish diaspora embraces a different sense of place, unrelated to contemporary cultural networks that connect the physical space of Iraq to its London diaspora. Yet what remains is an embodied form of culinary nostalgia for the lost city.

Popularising *amba* in Israel

Unlike the proud but demographically humble Iraqi community in London, Israel saw the arrival of more than 100,000 immigrants in the 1950s. Here, *amba* was first confined to the Iraqi-Jewish urban enclaves of Ramat Gan and

Or Yehuda. However, the history of the experience of the Iraqi Jews in Israel was worlds apart from that of their London counterparts. Facing systematic discrimination on arrival, many spent years in poor, cramped *ma'abarot* (transit camps), though not without resistance (Bashkin 2017). The distinct smell and taste of Iraqi food often evoked disgust in Ashkenazi Jews and became a vessel for gastro-racism. Indeed, 'the immigrant body is culturally constructed to be the natural carrier and source of undesirable sensory experiences and is popularly perceived to be the site of polluting and negative olfactory signs' (Manalansan 2006: 41). Many *Mizrahi* (Eastern) Jews internalised an odorphobic sense of shame about their culture and ethnic food in the 1950s and 1960s, leading *amba* to remain confined to Iraqi diasporic spaces for decades. During the 1991 Gulf War, when some of Saddam Hussein's Scud missiles fell on Iraqi neighbourhoods in Ramat Gan, a popular joke was that those areas were targeted due to their strong smell of *amba*. In the film *Forget Baghdad* (2002), Ella Shohat recalls:

> One of the major ways that my parents, all the Iraqis around us, maintained their Iraqiness was by cooking Iraqi dishes all the time. I often feel like it was their way of sheltering themselves from a hostile world that was outside. . . . [T]hey were expected to bring lunch to school that imitated what Ashkenazim ate at the time. That was a roll with a chocolate spread. You know, that was something unheard of in my culture, but I was so ashamed, because what my mother cooked for me in school was the *beid 'im amba*, you know, an egg with *amba*. And that was horrible because it did smell, and one of my first traumas when I went to school was when I was called, *iraqit masreecha*, which was 'stinky Iraqi'. Yeah, and it stayed with me.

Such distinct memories of visceral shame are almost the polar opposite of the rich memories of culinary nostalgia felt by Iraqi Jews when reminiscing about their childhoods. Nonetheless, *amba* did remain an integral part of the Iraqi-Jewish diet, and indeed, the lack of availability of The Ship brand, which the Iraqi diaspora in London continued to consume, led to a new phenomenon: homemade *amba*. By this point, fenugreek had become an essential ingredient, perhaps in part due to the Yemenite-Iraqi urban cohabitation in *Mizrahi* neighbourhoods. *Amba* was now being produced at home and by a number of small-scale family enterprises for local Iraqi consumption. Today, in south Tel Aviv, the Hatikva market stalls serve homemade *kubbeh* with *amba*, and a delicatessen owned by the Ofer family – whose members claim that their grandfather was the first to bring *amba* to Israel – can be found alongside a shop selling delicacies, owned by the Syrian Amiga family (Figure 6.2). In Or Yehuda in central Israel, there is 'Little Iraq' that markets 'original Iraqi *amba*'.

In the 1960s, *amba* began its journey towards becoming an everyday condiment simultaneously loved and abhorred by Israelis. Mr. Sabih Halabi and his wife opened a small kiosk in Ramat Gan, which for the first time marketed

Figure 6.2 Chunks of mango in thick *amba* at Hatikva market

Source: Photograph by Joel R. Hart

Shabbat food to bus drivers on weekdays. Through our research, we learned that since then, Israelis associate the dish with Mr. Sabih, who distributed it beyond the Iraqi family circle. In the 1980s, *amba* became, like Tunisian *harissa* and Yemenite *zhoug*, another condiment that adds a distinctive spicy note to pitas filled with shawarma, falafel, and fried aubergine. This process, which we call the 'falafelisation of the sabih', established the status of *amba* as a key component of pita bread–based street food in Israel and beyond.

The gradual popularisation of *amba* beyond the bounds of the Iraqi community positioned it as an all-Israeli or Palestinian 'sauce' consumed equally by Mizrahi, Ashkenazi, and Palestinian street foodies (Monterescu and Hart 2018; Shams 2018). For Palestinian chef Salah Kurdi, Israeliness is nothing but simple. In his Al-Ashi restaurant, he seeks to salvage Palestinian cuisine using indigenous recipes from the port city of Jaffa. During our conversation, he told us:

> I was once asked how Israeliness has affected Palestinian cuisine. I replied that this happened when we started seeing hummus in the home refrigerator. Even though it's a national Palestinian food we were exposed to it only once a week, on Saturdays at the local hummus eatery.

In contrast to hummus as the national icon, Kurdi states,

> *Amba* rolled into our lives but no one is claiming ownership over it. I can't say that when I cook Arab Palestinian dishes *amba* is an authentic component. *Amba* is my Israeliness.

Since the 1990s, *amba* has become so widespread that every Israeli has a strong opinion on it. Due to its circulation and popularisation, this fervour is not connected to projects of nation-building nor reducible to ethnic identification. Rather, it has grown larger than being associated with one ethnicity or nation. With the 'falafelisation of sabih', it is no longer only an Iraqi-immigrant food but a condiment that Israelis love or hate, dividing Ashkenazi and Mizrahi, men and women alike.

Clandestine consumption: from ethnicity to gender

Despite *amba*'s growing popularity, a persistent discourse never ceases to highlight its contentious smell. Like alcohol, *amba* is 'embodied material culture' (Dietler 2006), which blurs the boundaries between inside and outside, private and public. While in Iraq *amba* consumption was not gendered, in Israel it indexes gender preferences as it remains anchored in the country's ethnicised foodscape. Its place on a pedestal of male food consumption probably derives from its sour taste and pungent smell, denoting a supposedly masculine ability to handle rougher, more intense flavours and odours. In a widely read article in one of Israel's national newspapers, titled 'An embarrassing radius of stink', an *amba* aficionado confessed he is 'deeply in love with *amba*' (Dor 2011). 'Who doesn't know the wonderful feeling you have when you sense the tanginess of this divine liquid that turns any anaemic schnitzel to real gastronomy? It's the world's best sauce', he declares. The excited writer then goes on to describe his dangerous liaisons with *amba*, which amount to no less than conjugal unfaithfulness:

> I was unfaithful to the queen, who is my wife, minutes after she boarded the plane for the simple reason that my affair with the yellow mistress cannot exist on an everyday basis. *Amba* has a tendency to exit your body days after your intimate encounter from every possible pore in your body and in such a smelly way that you are surrounded by an embarrassing radius and no one would dare to approach you, not even mosquitos.
>
> (Dor 2011)

He then reveals that because *amba*'s 'smell is so unbearable to my wife . . . she put an absolute veto on my encounters'. Amounting to an exchange of bodily fluids, the writer relays his negotiations over the terms of encounter:

> A hot session with my darling *amba* can happen only when my wife and I are hundreds of miles apart . . . I allow you to eat *amba*, my wife warned

> me when she checked her boarding cards and passport, but 48 hours before my arrival, you cannot touch it. You will be sorry if I ever find it in the fridge!

The distressed author concludes with a plea to the public:

> Why isn't there a way for me to reach nirvana and eat *amba* without paying the price for it days after? Please help! Any tip will make me happy and less stinky.

One reader responded online to this distressed call by labelling the writer a 'clone of my husband' and added there is no solution for you 'but to move to India'. Another female reader confessed 'it stinks like a skunk' and added 'I smell pheromones of *amba* with fenugreek in the air more than half an hour after a Yemenite, Indian or Iraqi cross the street' (see talkbacks Dor 2011).

Other testimonies of gendered *amba* odorphobia flood Israeli public discourse. In one newspaper article titled 'Use perfume instead of *amba* and then we'll see' (Ferrara 2010), the female writer recounts the chronicles of her failed blind date:

> We walked into a pub and ordered something to drink. I gave him a second look. Perhaps it's not so bad as it seemed. Perhaps the date will be successful after all, only that he smells really bad! Am I losing my mind? Am I hallucinating? Can it be *amba*? I was flooded with anger and decided to ask a direct question: 'Tell me, do you usually show up like that to first dates with girls?' He stopped in the middle of a bite: 'Yes. I like to be myself on dates. I like to feel free'. I left him right there and asked myself how would he feel if I showed up after eating lots of fresh garlic.
>
> (Ferrara 2010)

The writer concluded her '*amba* tragedy' with the statement 'this is not a hate column against men' (ibid). The gendered nature of food consumption in Israel, however, calls to position *amba* in a larger framework. The gendered consumption of hummus (Hirsch 2016; Talshir 2019) and the public display of *mangal* (barbecue), particularly during Israel's Independence Day (Avieli 2017), demonstrate the importance of food practices in the expression of masculine power in Israel. Premised on the Zionist construction of the New Jew, free from the shackles of diasporic subjugation, the territorialisation of foods and the public affiliation of food practices with the Israeli flag are commonplace. Indeed, 'there is perhaps nothing more closely associated with masculinity in Israel than hummus' (Talshir 2019), and as Hirsch (2016) notes, 'masculine cultural capital' emerges from public display of hummus knowledge and the discussion of best hummus joints.

Amba, however, illustrates a more ambiguous social appropriation, public and private at the same time. During one day at Mr. Sabih's street food stall in

Ramat Gan, we counted some 30 male and 20 female customers. While the clientele was of mixed ethnic background, only one woman requested *amba* in her pita. She decried, 'I love it for the taste, but hate it for the smell'. *Amba*'s distinctly gendered 'pheromones' are symbiotic with its status as a diasporic food. *Amba* lovers cling on to its very material essence, its sensoria, rather than imbuing it with divisive political meanings. Most Israelis see hummus, like the land, as 'sacred' and lauded as such: its very colour, 'blonde, like the landscapes of the region' (Caland 2018), evokes nationalistic desires to assert authenticity and control of the wild desert (Zerubavel 2019). Conversely, *amba* can evoke strong feelings of aversion. Seen by many as harsh, sour, and coarse and defined by its powerful aroma rather than taste and sight, it remains beyond the connection of food to the symbolic landscapes of Israel/Palestine. Whilst it certainly is the object of gendered discourses, such processes are often framed as private, even clandestine affairs. As Manalansan writes, 'As a social marker that evokes change and liminality, smell therefore provides a strategic mode of communicating identities, bodies and temporalities that are betwixt and between' (2006: 42). *Amba*, now Israelised but always evoking the scent of elsewhere, is a matter of strong but individual preference, exhibiting the diasporic and liminal even within gendered forms of consumption.

Migratory foods, sensory cities

This chapter followed the thing known as *amba* in its tortuous trajectories across the Indian Ocean from India to Iraq and from there to Israel, Palestine, and the United Kingdom. Consumed by Jews and Palestinians alike in Israel and by Iraqi Londoners and London foodies, *amba*'s distinct taste and aromatic persistence generates deep forms of urban memory and contentious ethnic and gendered discourses. Dodging politicisation (such as in the Hummus Wars), *amba* challenges notions of gastronationalism and cosmopolitanism alike.

From a sensory perspective, *amba* reconfigures the relationships between memory, city, and belonging. Following *amba* highlights its ability to cross borders and its ambiguous character, yet a sensorial analysis points to its recurring features of attraction and repulsion, sweat and smell. Starting as a domestic culinary marker of ethnic belonging and a mnemonic device for culinary nostalgia amongst Iraqi Jews, it gradually transcended the boundaries of the ethnic enclave and assumed a central yet contentious and gendered place at the heart of Israeli and Palestinian street food. Its mobility has led to multiple, overlapping, and sometimes contradictory associations with distinctive places of origin. Indeed, from the outset, the dissemination of *amba* has been a process of community-making and transcultural exchange centred around urban networks. From colonial Bombay to Basra, from Baghdad to Ramat Gan, from the Middle East to London, migrants settled in cities where *amba* articulates personal biographies, collective memory, and legacies of cultural syncretism. Always at a safe distance from gastro-essentialism and territorial politics, it indexes the migratory power of foods and their sensorial traces in cities.

The fragmented continuities between urban networks invoked by culinary nostalgia, taste, and smell offer an alternative reading of place. The articulations of place as sensorial mappings call for a relational understanding of food and the geography of city space. The urban diasporisation of Jewish communities and Iraqi commodities facilitated a heightened engagement with the senses and produced in the process new forms of sensorial proclivities – with interpretations ranging from ethnic odorphobia to male fascination and women's repugnance with *amba*'s bodily effects. As a space of mediation, the city allowed for the dissemination of foreign food to new consumers as it reproduced a lingering nostalgic affect among generations of migrants who seek to reconcile their cherished heritage with their current life away from the homeland.

References

Appadurai A. (1986) *The Social Life of Things: Commodities in Cultural Perspective*. Cambridge: Cambridge University Press.

Ariel A. (2019) 'Mosaic or Melting Pot: The Transformation of Middle Eastern Jewish Foodways in Israel', in Diner H. R. and Cinotto S. (eds) *Global Jewish Foodways: A History*. Lincoln: University of Nebraska Press: 91–114.

Avieli N. (2017) *Food and Power: A Culinary Ethnography of Israel*. Berkeley: University of California Press.

Bashkin O. (2012) *New Babylonians: A History of Jews in Modern Iraq*. Stanford, CA: Stanford University Press.

Bashkin O. (2017) *Impossible Exodus: Iraqi Jews in Israel*. Stanford, CA: Stanford University Press.

Berg N. E. (2019) 'Jews Among Muslims: Culinary Contexts', in Diner H. R. and Cinotto S. (eds) *Global Jewish Foodways: A History*. Lincoln: University of Nebraska Press: 70–90.

Boym S. (2001) *The Future of Nostalgia*. New York: Basic Books.

Caland B. (2018) 'Hummus, the International Life of an Indigenous Food', Presentation at *Food and Society International Conference: Food and Indigeneity*, Paris, 29–30 March.

Cook I. *et al.* (2004) 'Follow the Thing: Papaya', *Antipode* 36 (4): 642–664.

Dangoor L. (2011) *Flavours of Babylon: A Family Cookbook*. London: Waterside Press.

Dietler M. (2006) 'Alcohol: Anthropological/Archaeological Perspectives', *Annual Review of Anthropology* 35 (1): 229–249.

Dor M. (2011) (Hebrew) '*Amba*: A Week with My Lover', *Ynet*, 12 August, www.ynet.co.il/articles/0,7340,L-4103507,00.html.

Fabian J. (2007) *Memory Against Culture: Arguments and Reminders*. Durham and London: Duke University Press.

Ferrara E. (2010) 'Use Perfume Instead of Amba and Then We'll See', *Ynet*, 28 May, www.ynet.co.il/articles/0,7340,L-3893535,00.html.

Hart J. R. (2016) 'Between Baghdad and Babylon: Diaspora, Memory and Identity Amongst Iraqi Jews in London', Unpublished MA Thesis, The University of Cambridge, Cambridge.

Hirsch D. (2016) 'Hummus Masculinity in Israel', *Food, Culture, & Society: An International Journal of Multidisciplinary Research* 19 (2): 337–359.

Loizos P. (1999) *Ottoman Half-Lives: Long-Term Perspectives on Particular Forced Migrations. Colson Lecture 1999*. Oxford: Oxford University Press.

Manalansan IV M. F. (2006) 'Immigrant Lives and the Politics of Olfaction in the Global City', in Drobnik J. (ed) *The Smell Culture Reader (Sensory Formations)*. Oxford and New York: Berg.

Monterescu D. and Hart J. (2018) 'The Mango Sauce Connecting Indians, Israelis, and Palestinians and Taking High End Restaurants by Storm', *Haaretz*, Tel Aviv, 1 June.

Rhys-Taylor A. (2014) 'Intersemiotic Fruit: Mangoes, Multiculture, and the City', in Jones H. and Jackson E. (eds) *Stories of Cosmopolitan Belonging: Emotion and Location*. Abindgon: Routledge: 44–56.

Samir S. (Director) (2002) *Forget Baghdad: Jews and Arabs – the Iraqi Connection*. Germany and Switzerland: Dschoint Ventschr Filmproduktion AG.

Shams A. (2018) 'Amba: The Iraqi Mango Pickle from India That Tops Palestinian Shawarma', *Ajam Media Collective*, 22 October.

Somekh S. (2007) *Baghdad, Yesterday: The Making of an Arab Jew*. Jerusalem: Ibis Editions.

Swislocki M. (2008) *Culinary Nostalgia: Regional Food Culture and the Urban Experience in Shanghai*. Stanford, CA: Stanford University Press.

Talshir U. (2019) 'From Mundane to Macho: In Israel, Hummus Makes the Man', *Haaretz*, Tel Aviv, 1 January, www.haaretz.co.il/gallery/.premium-1.6785141.

Tsing A. L. (2015) *The Mushroom at the End of the World: On the Possibility of Life in Capitalist Ruins*. Princeton, NJ: Princeton University Press.

Zerubavel Y. (2019) *Desert in the Promised Land*. Stanford, CA: Stanford University Press.

7 Thuringian festive cakes

Women's labour of love and the taste of *Heimat*

Grit Wesser

Having lived well over a decade abroad, I returned in November 2012 to my parental home, located in a small Thuringian village in eastern Germany, for fieldwork. On the first Sunday after my arrival in the *Heimat* (home, native region), my mother announced that she would bake a white buttercream torte. It is neither my favourite cake, nor had she actually ever made this cake before – and so her choice perplexed me. However, I quickly ascribed her preference to the fact that November was not the right season for baking one of my favourites – an apple or a plum tray cake – as she would not be able to use fresh fruit from the trees in our garden. A whiff of sweetness still lingered in the entire house when she presented the cake to my father and me in the afternoon. He critically analysed every bit of it: the texture of the buttercream, the amount of strawberry jam used, the consistency of the sponge. The cake did not look quite as perfect, but I instantly recognised it as the same type of cake my late grandmother's baker was famous for far beyond his village. Tasting this special combination – the richness of the buttercream, the sweetness of my mum's homemade strawberry jam, the fluffiness of the sponge layers – immediately brought back all the warm feel-good moments of my childhood, when my extended family would sit around my maternal grandmother's living room table on some festive occasion. It was this very Proustian moment of involuntary memory, which – despite including many absent family members – was so pleasant that made me realise that my mother expressed her love through homemade cake and that my being with them for a year was a kind of special – almost unimaginable – family reunion.

Because of their sweetness and status as a treat, cakes are great social conductors and frequently feature at festivities in many places worldwide, yet which cakes and how and with whom they are consumed are of anthropological interest in understanding social belonging (Bloch 2005). In Germany, as Satsuki Sakuragi (2008) in her study of the German baking culture shows, until the 1930s, cakes were regarded a luxury as to make them required expensive ingredients, especially sugar and butter, and also particular kitchen appliances and culinary expertise. As such, only aristocracy and the upper middle class could afford them regularly while the majority of Germans enjoyed cakes solely at special events, such as life-cycle celebrations or seasonal festivals. The pertinent association of cakes with the familial home developed only at the beginning of the twentieth century – and in no small part due to strategic advertising campaigns by the German food company Dr Oetker, which aimed to increase its sale of baking powder (Sakuragi 2008: 62–87). Hailed as the economical alternative to buying cakes at the baker's, baking cakes at home was increasingly promoted as a virtue of the good bourgeois housewife. Such Sunday afternoon occasions as described in the introductory vignette, when family or friends consume coffee and cake together, are in Germany quite literally referred to as *Kaffee und Kuchen* (coffee and cake). This Sunday ritual, which associates the pleasure of eating cake with the cosiness and love of the harmonious family home, became an established practice only after West Germany experienced its 'economic miracle' in the 1950s. This prosperity enabled all social strata to afford regular cake consumption and was accompanied by the widespread view that the serving of homemade cakes was the 'dear duty' (*liebe Pflicht*) of a good housewife and mother (Sakuragi 2008: 86, 263–267).

This gendered discourse in which a woman's skills and labour are rendered inherent to her femininity – characterised by care and love – endures in most Euro-American societies (Cairns and Johnston 2015; DeVault 1991). In post-socialist societies, like the former East Germany (GDR/German Democratic Republic), this gendered discourse also persists, despite state feminism's attempt to break with gender ideologies that view women's place solely in the domestic sphere (for example, see Haukanes 2001). During the Cold War, the two German states' social policies 'mirror-imaged' each other: West Germany viewed women's role best reflected in the traditional trinity of *Kinder, Küche, Kirche* (children, kitchen, church) while East Germany promoted women as mothers and workers to free them from oppression and to fill labour force shortages (Borneman 1992; Weinreb 2017). Yet, as Alice Weinreb (2017) illustrates, despite these radically different approaches to women's labour, both states perceived feeding the family as women's 'natural' responsibility, linked to maternal love. Perhaps unsurprisingly, then, during my fieldwork Thuringian men and women used the common German proverb that 'love goes through the stomach' (*Liebe geht durch den Magen*) almost exclusively to assess a woman's cooking or baking skills for attaining a good spouse or for feeding her family appropriately.

In this chapter, I aim to go beyond exploring the role of homemade cakes as a material expression of maternal love and nurture of the familial home. Rather, my interest is in how such gendered work of creating emotional attachment is fundamental in transforming homemade cakes to Thuringian festive cakes at life-cycle rituals – and in this process serves to create connections to *Heimat*. It is through particular sensory qualities that facilitate convivial commensality that the familial home expands to the regional home through the intertwining processes of producing and consuming these festive cakes. Before I explore the gustatory, aesthetic, and tactile qualities that foster this particular kind of commensality, I briefly address another sense that profoundly depends on our sensory experience: the sense of belonging.

Heimat: the German sense of belonging

Commonly translated as either 'home', 'homeland', or 'native region', these English terms do not fully capture the multifarious connotations and emotional significance German-speaking people associate with *Heimat* (Applegate 1990; Blickle 2002). However, these translations hint at *Heimat*'s ambiguity: do Germans mean their familial home, the nation-state to which they have claims, or the region in which they were born and bred? While this question appears a matter of scalar senses of belonging, the meaning of *Heimat* has changed throughout history. Originally referring to the homestead and property, 'since the second half of the 18th century *Heimat* has become increasingly associated with an inner emotional capacity to attach oneself with personalized memories of experiences to a place, a family, a specific landscape' (Blickle 2002: 78) and, with the rise of the nation-state, also to one's nation. After 1945, *Heimat* – in opposition to its use by the Nazi regime – moved from a national to a local concept to enable West Germans to have pride in Germanness 'without associating it with the militarist, state-led nationalism of the Third Reich' (Confino 2006: 64). Today *Heimat* 'can refer to both local and national "homes" [it] evokes Germanness, while appearing distinct from the nation's tainted pasts' (James 2012: 6). *Heimat* is thus a spatial sense of belonging but, simultaneously, an imaginary of a harmonious, apolitical place that enables Germans to articulate feelings of loss, such as of one's childhood or a past sense of community, when everything seemed simpler because everyone knew their place in the social order (Blickle 2002; James 2012; Palmowski 2009). The majority of scholarship on *Heimat* focuses on either the analysis of film in the *Heimat* genre (von Moltke 2005) or local heritage associations as beacons for fostering such a spatial sense of belonging (Applegate 1990; James 2012; Palmowski 2009). In contrast, I see the family as a source and agent of (re)creating such ties of belonging beyond the familial home and the role of food and the senses in it.

While 'the sensual qualities of food evoke visceral responses that transform external, anonymous social processes into intimate, immediate, and personal experiences', as Melissa Caldwell (2009: 3) observes, my focus is on the opposite process. How can these sensual qualities of food evoke visceral responses that enable embedding intimate, immediate, and personal experiences into social processes of creating senses of belonging to a locality and imaginary nation? I take up the suggestion that feeding and practices of commensality can be scaled up from familial to regional home – with the potential to incorporate the nation – (Carsten 2004; see also Thelen and Alber 2018: 12–13) by exploring such processes and their sensuous, emotional, and gendered nature in the case of Thuringian festive cakes.

Fieldwork context

As a Thuringian myself, to me these cakes did not stand out at all – yet I certainly would have noticed their absence at large festivities. They became the object of ethnographic interest during my doctoral fieldwork in 2013 on the secular coming-of-age ritual *Jugendweihe* ('youth consecration'). This ritual continues to be associated with the former GDR and remains a significant milestone of the eastern German life-course. The east Thuringian city in which I conducted the greatest part of my fieldwork is a traditional worker's city with just under 100,000 inhabitants and was a centre of heavy industry during state socialism. With the collapse of state socialism in 1989–1990, its population suffered from high unemployment that – accompanied by outward migration – led to administrative reforms, which incorporated many of the city's adjacent villages in the 1990s. While these villages gained city status, their character remains rural, perhaps best categorised as peri-urban. As part of my fieldwork, I talked to and accompanied five families in preparing and celebrating *Jugendweihe* to trace the continuities and changes since the *Wende* (the political turn-around of 1989–1990). During this research, and during shorter subsequent visits in 2015, 2016, and 2018, I also attended for comparative reasons other secular and religious life-cycle celebrations in Thuringia. Here, I draw on observed practices and conversations with family members and friends present at festivities but especially with mothers, who are usually responsible for organising such large family celebrations.

These Thuringian families who celebrated their offspring's *Jugendweihe* or confirmation between 2012 and 2015 differ in terms of their educational background, economic standing, and political disposition, but they also share significant features. Parents have in common that they spent their formative years in the GDR, and grandparents that they spent most of their working life building that same country – a country that no longer exists. Both generations experienced German reunification as a displacement – echoed in the popular saying of having 'emigrated without leaving [home]' (Berdahl 1999: 202). Notably, none of my interlocutors wished for a return of the former GDR, but all expressed in one way or another that they did not feel they fully belonged to contemporary Germany. This prevailing perception of eastern Germans as second-class citizens has been rightly associated with the devaluation of East Germans' biographies and persisting socio-economic inequalities between the former East and West Germany (Kolbe 2010). Yet national belonging cannot be fully understood through focusing on loyalty to specific political ideals or social communities; we also need to explore people's 'attachment to particular tastes, smells, sounds and sights, which themselves carry cultural values and personal memories' (Howes and Classen 2014: 65). I contend that East Germans' feeling of having lost their home was also due to fundamental sensory changes thrust upon them with the arrival of a different political economy overnight. After the Berlin Wall fell, the majority of East Germans described their first visit to West Germany in terms of sensory overload. While they were excited to finally experience the 'Golden West' by themselves, they simultaneously lamented that there were too many lights, too many colours, too many odours, too much advertising, too much noise, too much choice. This 'too-much-of-everything' sense-scape, or intentionally designed *hyperesthesia* of consumer capitalism (Howes 2005: 281–303), stimulated the East Germans' initial shopping frenzy for long-desired Western products but soon also reshaped their own cities and hometowns beyond recognition. Set against this backdrop, in what follows, I show how discourses and practices surrounding Thuringian cakes are one way of recuperating a simpler but emotionally laden sense-scape, drawing on the familial home, to make Thuringians feel at home in Thuringia and, ultimately, in contemporary Germany.

From *Blechkuchen* to *Festtagskuchen*: making ordinary cakes festive

Germans take their bakery products seriously, a stance exemplified not only by my father's rigorous assessment of my mother's homemade cake but also by the many artisanal bakeries that mark German cityscapes. For Thuringians, *Kuchen* is an elevated form of bread (Braungart 1994: 7) – the nation's staple. My Thuringian interlocutors frequently bought bread either in artisan bakeries or in bakeries that were part of a supermarket chain. Even during state socialism, when East Germany's political elite believed that collectivisation and large-scale production would pave the way to communism, private bakeries accounted for more than 40 per cent of all baked goods (see Buechler and Buechler 1999: 801). Although my interlocutors also bought cakes from bakeries, especially in rural areas, it is still very common to bake them at home on a weekend. Sunday coffee and cake tends to include freshly brewed coffee served with one or perhaps two different types of usually homemade cake. Unlike the torte my mother made, *Blechkuchen* or tray cakes – such as the plum or apple cake I had hoped for – are the most frequently baked cakes. *Blechkuchen* is an umbrella term for a great variety of cakes that are baked in a large rectangular tray instead of a round baking tin (see Figure 7.1). These cakes usually have a solid dough base, either *Hefekuchen* (yeast cake), *Biskuit* (sponge), or *Mürbeteig* (short pastry) and are served in squares not unlike a brownie.

Kaffee und Kuchen features at all major life-cycle celebrations, where good hosts offer them to their guests, but this aspect of the celebration differs from Sunday coffee and cake in two significant ways. Firstly, the table is set in a more formal manner, including a tablecloth, a (matching and perhaps more precious) porcelain coffee set, flower decorations, candles, and place cards to create a festive atmosphere. Secondly, while hosts often offer two or three round cakes, cut into eight or twelve triangular slices, they are not an essential element of the festivity. Rather, to make any Thuringian life-cycle celebration successful, it is crucial to have a great variety of Thuringian *Festtagskuchen* (festive cakes) on offer. These festive cakes are essentially a selection – never less than ten different types – of tray cakes. Beside the festive decoration, this wide range makes coffee and cake special as one grandmother pointed out to me: 'The more types of [tray] cakes, the more festive the occasion!' So what then makes these cakes 'Thuringian'?

Figure 7.1 Thuringian plum tray cake
Source: Photograph by Grit Wesser

Tasting and presenting Thuringia

The great selection of tray cakes, as Thuringians of all ages explained to me, ensured that there was something suited to everyone's taste. Thuringians differentiate between dry (*trocken*) and wet (*nassen*) *Kuchen*, and the variety of cakes could comprise, for example, white or brown *Streuselkuchen* (streusel or crumble cake), *Bienenstich* ('bee sting' – almond and honey cake), and nut cakes for those who prefer dry cakes. For those who prefer wet cakes, apple, plum, cherry, tangerine, or a combination of tropical fruit cakes are on offer. There are also rhubarb, gooseberry, or red current cakes available for the more sour/tart taste buds and different types of buttercream as well as *Mohnkuchen* (poppy seed cake) for those who prefer creamier options. Frequently, there would be a version of *Papageikuchen* ('parrot cake', consisting of at least three differently coloured parts) that children particularly favoured. Other popular variations were *Eierlikörkuchen* (German eggnog cake), *'Eiskrem'-Kuchen* ('ice cream' cake), and several types of *Quarkkuchen* (a lighter version of a British baked cheesecake).

Not only the taste and colourful selection matter, but also the presentation of the cakes, which differed from Sunday *Kaffee und Kuchen*. While it was common to cut *Blechkuchen* into squares or rectangles, for major festivities,

Thuringians cut the same cake with more precision into smaller matchbox-size pieces. These pieces, one or two mouthfuls each, of at least five to ten different types of tray cakes were nicely arranged and offered on a platter (see Figure 7.2). This way of serving cakes is a regional custom, a fact that I only began to register when a guest – a German woman from a different region – disapprovingly remarked on it. She deemed both the size and the shape of the servings 'inappropriate': a piece of cake had to be triangular and of a size that you could not eat more than two without feeling ill. For Thuringians, on the contrary, it was essential to present their guests with a great selection of cakes. Because there is such a great variety of cakes that guests might be disappointed about being unable to try them all, the small size of the pieces was, as most Thuringians elaborated, a host's thoughtful measure: it enabled guests to taste all types of cakes, if they wished.

Indeed, the shift from a standard-size piece of tray cake to bite-size cut pieces of cakes for festivities is crucial in creating convivial commensality that differs from Sunday coffee and cake: namely, in the larger number of people consuming cakes together. In the remainder of this chapter, I will explore the intertwined processes of producing and consuming these cakes in regards to the senses and to social belonging.

Figure 7.2 Thuringian festive cakes

Source: Photograph by Grit Wesser

The gendered labour of love and the Thuringian 'baking woman'

The making of cakes has always been the business of women, yet it has not always included the baking of cakes by a woman or at home. In the nineteenth century, for big life-cycle events, such as weddings, cakes would be baked either at home (homesteads had their own large baking ovens) or in a communal baking house, where women from the entire village would help (Koch and Sander 2014: 28). As one grandmother explained, in the pre-socialist past, it was common for a wealthier farmer to employ a woman who would bake the cakes for such a big festivity. For less well-off families, female family members would share the laborious task of making cakes, carry or cart them to a nearby local bakery for baking, and pick them up the following day – a practice that continued into the early 1980s in rural Thuringia (see also Koch and Sander 2014: 32). Hence, for Thuringians to consider a cake 'homemade' does not require the baking of the cake at home but is reliant on women's entire production process before baking as well as the cutting and presentation of the baked cakes.

For virtually all life-cycle rituals, Thuringians frown upon the buying of cakes in a supermarket. On one occasion, when a mother also served two bought cakes, she pointed it out immediately before any guest could potentially raise a complaint. She explained that the cakes were of a particular well-established brand and had been tested by her and her family beforehand and found to be of an acceptable standard – that is, homemade-like. The usage of '*gekaufter Kuchen*' (bought cakes) was somewhat misleading because it enabled a distinction from cakes that hosts had, in fact, also bought but regarded as '*hausgemacht*' (homemade) – just not necessarily in their own home.[1] Most hosts, usually full-time working mothers, do not have the time, the facilities, the culinary skills, and/or the desire to bake such a great variety of cakes. As one mother ironically replied when I teasingly asked her whether she had baked all the cakes for a festivity herself: 'Sure, Grit – the whole of last week because I had nothing else to do!'

Instead of baking themselves or buying cakes from a bakery, for my Thuringian interlocutors, it was customary to arrange with a so-called *Backfrau* (baking woman) to bake a selection of cakes for large life-cycle events. The phenomenon of the *Backfrau* seems to have developed during socialist times, when – unlike in West Germany – most women already worked full time outside the home but nevertheless also bore the brunt of domestic chores – a double burden (Weinreb 2017). Mothers, who tend to be in charge of organising big festivities such as *Jugendweihe*, would provide a *Backfrau*, who lived in the same or a neighbouring village, with hard-to-get ingredients so that she could bake an assortment of cakes. 'Baking women' commonly acquired their skills through helping their mothers bake at home when they were still girls and by continued practice rather than through professional training leading to a qualification as a baker or pastry chef. Today, they tend to run small (official) baking businesses or to be friends or acquaintances who would bake specially for such big occasions by way of earning an extra (usually unreported/untaxed) income. Both types of baking women rely largely on word of mouth spread through contented customers.

While nowadays, it is no longer necessary to provide ingredients, families usually pick up the cakes – like during socialist times – on the day of the celebration or the previous afternoon/evening.[2] Such arrangements had to be made in good time prior to the party in order to decide on the selection and also – in the case of a 'baking woman' whose services one had not previously used – in order to taste some of her products to be certain of her baking skills. One mother explained to me that she used the same baking woman for her youngest son's *Jugendweihe* as she had used for her older children's coming-of-age celebrations. She allowed each of her three children to pick one or two of their favourite types of cakes before she decided and ordered the rest of the cake assortment.

The preference for using a baking woman instead of buying cakes from a baker, I suggest, is not only for economic reasons but highlights the association of home, love, and femininity for achieving the right quality and sentiment of cakes. East German bakeries' labour division was gendered: men tended to be bakers and pastry chefs while women often only sold bakery products (Buechler and Buechler 1999) – a distinction that largely persists. At home, women generally make cakes, and these cakes are associated with mothers or grandmothers and their 'natural' role as familial nurturers. This notion of women's natural capability to create cakes and of men as having to cultivate baking skills as a trade is perhaps best exemplified by a (male) Thuringian pastry chef – quoted in a regional newspaper – summarising the Thuringian passion for baking as 'cake is love, torte is artisanry' (Glase 2018). Unlike the suggestion of the pastry chef, however, a *Backfrau*'s baking skills are paramount. Although the Thuringian *Blechkuchen* or *Festtagskuchen* are rather simple compared to, for example, an elaborate, intricately decorated three-tiered wedding torte, a *Backfrau* not only needs to bake delicious cakes; she also has to excel at variety, precision, and efficiency.

Baking women tend to offer 30 to 50 different types of cakes and are unlikely to share their recipes as is otherwise common among friends and family members. In the thirtieth edition of what may be the most popular Thuringian baking recipe book, the retired Thuringian *Backfrau* Gudrun Dietze explains that 'our cakes do without expensive ingredients. We have already learned from our mothers to respect the simple things' (Dietze 2013: 8). Baking Thuringian festive cakes requires the skill to create diversity with few and basic ingredients as well as precision because – despite different textures and elements – the cakes should all be about the same height and never be higher than two centimetres. This particular attention to shape and size enables consumers to pick up a piece of cake easily either with a cake fork or with their hands, holding it between thumb and index finger and devouring it in one or two bites. As a result, baking women need to consider precision for the cake's presentation of bite-size pieces that enable consumers to have a tactile experience with the cake and thus a more sensuous eating experience. For baking women to bake such a great assortment of cakes takes a lot of time and effort, and, as Dietze adds, for a successful outcome, '[t]he most important ingredients are patience and love' (ibid).

The use of a *Backfrau*, then, is a strategic way for Thuringian mothers to outsource their unpaid labour – turning it into paid labour – while simultaneously retaining the right quality and sentiment of 'homemade' cakes. It offers mothers a moral respite from having to live up to being the 'perfect mother'. Certainly, Thuringian bakers and pastry chefs have the culinary knowledge and skills to bake traditional Thuringian cakes, but the crucial ingredient for homemade-quality cakes is 'love'. Thuringians commonly ascribe this ability 'to make cakes with love' to women only. Since the requirement for Thuringian festive cakes is not that Thuringian mothers make them in their own home but that a Thuringian woman makes them, the feelings of home and love are projected onto the regional home. This gendered process of preparing cakes thus connects to and makes the cakes taste of *Heimat*: that is, the native region of Thuringia.

Consuming festive cakes

On the day of a life-cycle ritual, depending on the time of the public ceremony, the festivities will follow in such a way that coffee and cake may be the first or the second festive meal. Festive coffee and cake in Thuringia is – like Sunday coffee and cake – served between 3:00 and 4:00pm. Presenting one's guests with one particular cake, such as a wedding cake, is not a standard practice for life-cycle celebrations in Thuringia. Although wedding cakes have become increasingly indispensable at Thuringian weddings as well, this is a comparatively recent development. One grandmother explained to me that she and her husband did not even have a torte at their wedding in the 1960s – tortes, she insisted, were not at all common in the past.[3] In his study of British wedding cakes, Simon Charsley (1997) notes that, although his interlocutors viewed the three-layered white wedding cake as an essential element of a wedding, they did not particularly enjoy eating it: that is, the cake's creative beauty frequently trumped its taste (1997: 51). In contrast, in Thuringia even today, a wedding cake should not necessarily be of the typical three-layered and white kind but can be a relatively simple one-layer torte, which needs to please the wedding party through its taste. This emphasis on taste also holds true for the festive cakes. However, this does not mean that the cakes' presentation is inconsequential. In fact, the great variety of festive cakes also ensure that their presentation is colourful since, as the German saying goes, '*das Auge isst mit*' – it should also be a feast for the eyes. Rather, a cake is not supposed to be reduced to aesthetics, nor should its appearance infringe on or diminish its gustatory qualities.

Guests frequently commented on the cakes on offer and suggested to others present which type of cake to try next since they had judged it to be 'delicious' (*köstlich*), 'tasty' (*lecker*) or 'almost as good as grandma's' (*fast so gut wie Oma's*).

Mothers at coming-of-age celebrations, who were often anxious about making this one-off event special for their loved ones, would ask guests if such a gustatory approval was not given voluntarily. It was the ultimate compliment – and a sign of successful hospitality – when guests referred to the cakes' flavour positively by linking it to their grandmothers' baking. When I pondered aloud how curious it was that grandmothers' baking and not mothers' was seemingly the greatest praise, one grandmother volunteered her thoughts on the matter. She believed that because retired grandmothers had more time than working mothers, this positively affected the preparation and consumption of cakes because grandmothers had 'more time to care and love', which made the cakes taste better to their grandchildren. Often in their assessment of a cake, Thuringians referred not only to the way it looked, its right consistency, and its texture but also to the fact that the cakes were not too sweet. Particularly Thuringians of the grandparental and parental generations, who had developed their taste preferences during state socialism, would comment on this gustatory preference by describing store-bought cakes as 'too' or even 'disgustingly' sweet compared to homemade cakes. As a result, what Amy Trubek defined as fundamental to the French foodview – that eating and drinking 'needs to be a shared experience that incorporates sensory analysis *and* sensory pleasure' (Trubek 2008: 46) – also holds true for Thuringians.

The wealth of selection underlined the significance of the event, but perhaps more importantly, such a varied, well-presented range of cakes cut into small appetising pieces guaranteed that people ate together. Living and eating together – that is, sharing substance – are essential social processes for creating kin (Carsten 1995). Here the commensality at life-cycle events is a particular heightened form of embodied memory through seeing, feeling, tasting, evaluating, and ingesting cakes with others, as the triggering of my involuntary memory described earlier – sharing buttercream torte with my parents – attests. This particular kind of commensality reminded party attendees that they shared not only food with their commensals but also a family history punctuated by such family celebrations (Gillis 1997: 93). The variety of cakes and their presentation style reflect the ideal image of Thuringian hospitality: hosts are not only aware of people's varying gustatory preferences but accommodate them. This also ensured, I suggest, that there was a greater likelihood for guests to find at least one cake that reminded them of home or their grandmother's baking.

This sharing of food at the festivity, however, was extended in time and space in which cakes served to expand the familial home to the regional home by giving cakes to other extra-familial Thuringians beyond the family celebration. All of my interlocutors from the grandparental and parental generations recalled the custom of giving a *Kuchenpaket* or *Kuchenteller* – a little package or plate with a selection of the bite-size cut cakes – to gift-givers, usually neighbours and acquaintances, who did not attend the actual celebration. My interlocutors viewed this practice as an appropriate gesture of gratitude for the gifts received. During socialist times, as well as in the pre-socialist era, this practice served to expand the household in a network of mutual obligations with other households within the community. Gediminas Lankauskas (2015) identifies *šakotis*, the branchy cake – a traditional Lithuanian cake – as a key comestible at wedding celebrations in post-Soviet Lithuania, where guests consume this cake either before or after the final wedding toast, but departing guests also take parts of *šakotis* to their own homes. Here, the cake turns into an alimentary reminder of the wedding to be shared with others who did not attend the wedding – 'promoting it as a celebration centrally concerned with the continuity of the family, tradition, and Lithuanianness' (Lankauskas 2015: 210). In the German context, there is no one such cake that could successfully promote 'Germanness'. Instead, like West Germany in the 1950s and 60s, Thuringians recreate *Heimat* through culinary traditions that express a regional sense of belonging.

Today there is a noticeable decline in giving away festive cakes to gift-givers, especially in the urban centre. One mother, who grew up in the city centre but now lived with her family in a village, explained to me, 'You can't get away without cakes in a village!' In contrast, a family who lived in the city centre celebrated their son's *Jugendweihe* without having *Kaffee und Kuchen*, partly for economic and practical reasons of time. Both mother and grandmother recalled the practice of given away *Kuchenteller* for their own *Jugendweihe* celebrations, and the grandmother then commented: 'Well, back then, this street was just like a village community'. Consequently, the decline in this practice cannot simply be reduced to an urban/rural divide but reflects a change in social relations associated with changes in the political economy in the early 1990s, which for some families meant a retreat to, and a greater reliance on, their own household.

Conclusion

The example of Thuringian festive cakes shows that, in addition to the senses, we need to investigate their intersections with food, kinship, and gender to better comprehend processes of creating senses of belonging. I have demonstrated that the gendered nature and emotional significance of practices of feeding and the commensality of homemade cakes, which create a sense of familial belonging, are used by Thuringians to project such sensations to the regional home or *Heimat*. Thuringians achieved such expansion from familial to regional home through the skilful baking and presenting of Thuringian festive cakes at life-cycle celebrations that facilitated convivial commensality and simultaneously ensured sensory recalls of the familial home focused on taste and texture. Notably, unlike in the opening, where the smell of a freshly baked cake was essential to recall kinship, olfaction plays a secondary role in recalling *Heimat*. While Thuringian festive cakes promote familial continuity and tradition, they do so not despite but because of the political rupture of 1989–1990 – a rupture that most East Germans also associated with a loss of belonging. These cakes – unlike Lithuanians' *sakotis* – do not explicitly make claims to a national home – neither German nor East German – but to the native region of Thuringia as an 'untainted home' removed from the political pasts of two dictatorships. Thuringians' emphasis on simplicity, reflected in the use of 'simple ingredients' and the 'homemade', are also sensory reminders of a lost GDR sense-scape. At the same time, the sophistication of festive cakes mirrored in their gustatory variety, aesthetics, and presentation of bite-size pieces, negotiates their rightful place in a contemporary 'too-much-of-everything' sense-scape of neoliberal Germany.

Notes

1 Thuringians use both '*selbstgebackener*' (self-baked) and '*hausgebackener*' (home-baked) as well as '*hausgemachter*' (homemade) to refer to traditionally baked cakes.
2 Restaurants usually offer cakes from a local *Backfrau* for life-cycle celebrations too.
3 Germans differentiate between (wedding) *Torte* and (tray) *Kuchen*, both of which mean *cake* in English. Although the flavour focus for tortes is always their topping/filling, crucially, they are round in shape so that sometimes simple round *Kuchen* can also be called *Torte* (see also Sakuragi 2008: 22–29).

References

Applegate C. (1990) *A Nation of Provincials: The German Idea of Heimat*. Berkeley and Los Angeles: University of California Press.

Berdahl D. (1999) '"(N)Ostalgie" for the Present: Memory, Longing, and East German Things', *Ethnos* 64 (2): 192–211.

Blickle P. (2002) *Heimat: A Critical Theory of the German Idea of Homeland*. London: Camden House.

Bloch M. (2005) 'Commensality and Poisoning', in Bloch M. (ed) *Essays on Cultural Transmission*. Oxford: Berg: 45–59.

Borneman J. (1992) *Belonging in the Two Berlins: Kin, State, Nation*. Cambridge: Cambridge University Press.

Braungart M. (1994) *Backen in Thüringen*. Rudolstadt und Jena: Hain.

Buechler H. and Buechler J. M. (1999) 'The Bakers of Bernburg and the Logics of Communism and Capitalism', *American Ethnologist* 26 (4): 799–821.

Cairns K. and Johnston J. (2015) *Food and Femininity*. London and New York: Bloomsbury.

Caldwell M. L. (ed) (2009) *Food and Everyday Life in the Post-Socialist World*. Bloomington and Indianapolis: Indiana University Press.

Carsten J. (1995) 'The Substance of Kinship and the Heat of the Hearth: Feeding, Personhood, and Relatedness Among Malays in Pulau Langkawi', *American Ethnologist* 22 (2): 223–241.

Carsten J. (2004) *After Kinship*. Cambridge: Cambridge University Press.

Charsley S. (1997) 'Marriages, Weddings and Their Cakes', in Caplan P. (ed) *Food, Health and Identity*. London: Routledge: 50–70.

Confino A. (2006) *Germany as a Culture of Remembrance: Promises and Limits of Writing History*. Chapel Hill: University of North Carolina Press.

DeVault M. L. (1991) *Feeding the Family: The Social Organization of Caring as Gendered Work*. Chicago: University of Chicago Press.

Dietze G. (2013) *Thüringer Festtagskuchen*. Leipzig: Verlag für die Frau.

Gillis J. R. (1997) *A World of Their Own Making: A History of Myth and Ritual in Family Life*. Oxford: Oxford University Press.

Glase I. (2018) 'Thüringer Backzeit: Kuchen ist Liebe, Torte ist Handwerk', *Thüringer Allgemeine*, 22 December.

Haukanes H. (2001) 'Women as Nurturers: Food, Risk and Ideals of Care in the Czech Republic', in Haukanes H. (ed) *Women After Communism: Ideal Images and Real Lives*. Bergen: University of Bergen: 67–80.

Howes D. (2005) *Empire of the Senses: The Sensual Culture Reader*. Oxford and New York: Berg.

Howes D. and Classen C. (2014) *Ways of Sensing: Understanding the Senses in Society*. London and New York: Routledge.

James J. (2012) *Preservation and National Belonging in Eastern Germany: Heritage Fetishism and Redeeming Germanness*. London: Palgrave Macmillan.

Koch D. and Sander N. (2014) *Hochzeit in Thüringen: Trachten, Bräuche, Traditionen*. Erfurt: Sutton.

Kolbe M. (2010) 'Ostracism After Ostpolitik?: The "Second-Class Citizen" Phenomenon Among East Germans', MA Thesis, University of Georgia, Georgia.

Lankauskas G. (2015) *The Land of Weddings and Rain: Nation and Modernity in Post-Socialist Lithuania*. Toronto: University of Toronto Press.

Moltke J. von. (2005) *No Place Like Home: Locations of Heimat in German Cinema*. Berkeley: University of California Press.

Palmowski J. (2009) *Inventing a Socialist Nation: Heimat and the Politics of Everyday Life in the GDR, 1945–1990*. Cambridge: Cambridge University Press.

Sakuragi S. (2008) 'Vom Luxusgut zum Liebesbeweis: Zur sozialen Praxis und symbolischen Bedeutung des selbstgebackenen Kuchens', Unpublished PhD Thesis, Eberhard-Karls-Universität, Tübingen.

Thelen T. and Alber E. (eds) (2018) *Reconnecting State and Kinship*. Philadelphia: University of Pennsylvania Press.

Trubek A. B. (2008) *The Taste of Place: A Cultural Journey into Terroir*. Oakland: University of California Press.

Weinreb A. (2017) *Modern Hungers: Food and Power in Twentieth-Century Germany*. Oxford and New York: Oxford University Press.

8 The taste of home

Migrant foodscapes in marketplaces in Shantou, China

Shuhua Chen

We may sense home by sounds, smells, touches, sights, or tastes. Home, to this extent, is a multi-sensorial phenomenon experienced through the body. If home is considered 'the sensory world of everyday experience' (Ahmed 1999: 341), then migration may involve ruptures in the continuity of the sensory world since migrants may be distant from those senses. Can food smooth the ruptures and ease fragmentation or discontinuity? Can taste contribute to a nuanced appreciation of home in the context of migration? According to the National Bureau of Statistics of China (NBSC 2019), by 2018 the population of rural-urban migrants in China had reached 172,660,000. Most migrants return to their rural homes only once a year, usually during the Chinese New Year (Chan 2010). How does one experience home when being at home becomes one's 'annual holiday' (Chen 2018) while being away from home comes to stand for everyday life? Can a rural migrant ever feel at home in urban China?

Drawing on ethnographic fieldwork in Shantou in southern China, this chapter explores how rural migrants in urban China experience home in the context of food, ranging from buying to tasting, consuming to socialising. Through visual evocations of taste – a tour of a marketplace, where one can 'taste' ethnic mixing and different foodways – we gain a glimpse into various foodscapes and how locals and migrants price, shop for, identify, and experience food in the marketplace. I argue that, along with other practices in the urban setting, the rural migrants can experience home momentarily through everyday interpersonal interactions in purchasing food in marketplaces. It is through these interactions – physical as much as social – that the migrant experiences home as a sensory totality.

Why 'scape'? Industrialisation, urbanisation, and migrant foodscapes in Shantou

What do urban marketplaces reveal about foodscapes? How do these spaces contribute to understanding how a city and its dwellers are constantly being shaped and reshaped? What analytic value is added by considering 'foodscapes' rather than just 'food'? Why 'scape'? Geographer Kenneth Olwig traces the suffix -*scape* back to its origin:

> The suffixes -scape and -ship stem ultimately from an ancient Germanic root, spelt [*shape*] in modern English (basically a Germanic language). The power of this sense of shape lies in the dynamic relation between the meaning of shape as, on the one hand, an expression of -ship as an underlying nature, state or constitution which manifests itself through an active, creative, shaping process and, on the other, the material form which that process generates – its shape.
>
> (Olwig 2005: 21)

There are two aspects to explore with regard to a *scape*: the process of *being scaped* (shaped) and the *scape* (shape) itself. The aspect of how it is *scaped* emphasises the lived experience of being that shapes it: that is to say, *how we shape it* and, in turn, *how it shapes us*. Like two sides of a coin, the process of *being scaped* and the *scape* itself are entwined. It is essential to have embodied experience within the *scape* in order to explore the shaping process, as the 'distanced, contemplative and panoramic optic' (Ingold 2011: 126) towards it is by far not enough in understanding how a *scape* is shaped. If the shaping process were ignored, a *scape* would be lifeless (ibid). Hence, using *scape* to introduce the field – food in marketplaces in this case – ensures that, in addition to the usual sketch of the shape of the field included in ethnographies, we will not overlook and will draw attention to the fact that the field is not given but continually shaped. In this chapter, I will investigate the changing foodscapes in a suburban marketplace in the city of Shantou and how they have been shaped by migrant factory workers from other parts of China, as well as the negotiations behind the economic, social, and cultural coexistence of different foodscapes – the locals' and the migrants'.

An important historical port city with strong links to Chinese overseas and potential foreign investment, Shantou became, in the 1980s, one of the four original Special Economic Zone cities in China. The rise of light manufacturing in Shantou since the 1990s has attracted a large population of migrant workers who arrived from all over the country. From 2011 to 2013, I conducted fieldwork on rural-urban migration in Shantou, China, which involved working and living with migrant workers in a toy factory for fourteen months through participant observation. My fieldsite was Bomaqiao (the name 'Bomaqiao' is pseudonymised in order to protect interlocutors' privacy), a neighbourhood in Shantou that is undergoing rapid industrialisation and urbanisation. Bomaqiao, like many other suburban neighbourhoods in Shantou, has gone from being predominantly agricultural to being primarily manufacturing oriented in recent decades. Land is no longer used for farming; instead, a toy industrial park has arisen with more than a hundred small factories. In everyday life in Bomaqiao, I experienced the 'field' – that is to say, my 'sensual awareness of the smell, tastes, sounds and textures of life among the others' (Stoller 1997: 23) – and learned to be a 'tasteful fieldworker' (Stoller 1989: 29). Through the body's sensory entanglement in the life-world, I was able to inhabit Shantou's manufacturing landscape, breathe air that was polluted due to unchecked industrialisation, and taste its evolving foodscapes. The migrant foodscape in Bomaqiao's marketplace did not exist three decades ago. It came into being along with industrialisation and urbanisation, catering to the increasing inward migrant flow.

Experiencing foodscapes in marketplaces

Before the dramatic inward migration and economic liberalisation of the last three decades, the marketplace in Bomaqiao was an ordinary indoor market that sold local foods for a less-transient population. Now it has expanded in size to include two nearby parallel streets. Locals still entirely occupy all the stalls in the indoor market; stalls in the first street comprise a mixture of local and migrant proprietors while the second street caters entirely to migrants' favourite foods. Over the course of my fieldwork, whenever I walked from the indoor market to the first, mixed cuisine market street and thence to the migrants' market street, I would experience a changing soundscape: it would begin with the buzz of voices speaking only in the local dialect (Teochew), before becoming a mixture of Teochew and other dialects spoken by migrants. Eventually, I would find myself surrounded by voices without any trace of the local dialect. Just as the different soundscapes are shaped by the diversity of dialect and language, so, too, do kinds of food – their smells and the different ways in which sellers present them – fashion the different rhythms and choreographies of the indoor market and the two market streets, presenting us with two very distinct foodscapes – the locals' and the migrants'.

Visual evocations of taste: a tour of the marketplace

'I will show you next time when we buy food in the market', I often said to migrants with whom I worked when I found it difficult to put tastes into words or to convey a specific flavour of the local food in Shantou, which was new to most of them. As a scholar who grew up in Shantou in the 1980s and then returned to carry out fieldwork, I must have appeared both as someone who was intimately familiar with the local cultural landscape and also very much an outsider (initially at least) to the migrant population among whom I lived and worked. I will use a photographic narrative in an attempt to engage readers' other senses in a manner similar to the way I incorporated sight with other senses by bringing migrants to the markets. The photographic narrative focuses on the foodscapes in the Bomaqiao marketplace. Most photographs are from a shopping journey that I made with my interlocutor, Yang Cui, a migrant worker who comes from Sichuan province and who works in a toy factory in

Bomaqiao. During the journey, Yang Cui and I took photographs moment by moment of things that caught our eyes; each photograph represents a snapshot of a particular experience. Throughout the photographic narrative, I will reveal the cognitive processes behind those moments and how locals, migrants, and myself – in my double position of migrant and local – experienced food in Bomaqiao. As the photographic narrative progresses, it will compare the local market and the migrants' market, the practice of food consumption, and different experiences and 'tastes' of 'being at home' in the marketplace.

Foodscapes as a social and material phenomenon: 'But not today. . .'

It is about 6:30am. Yang Cui and I are walking towards the marketplace in Bomaqiao. Yang Cui prefers to do her daily food shopping in the early morning before the morning shift in the factory starts at 7:30am. We are walking towards the main street of the marketplace, which lies in between the local indoor market and the migrants' outdoor market street. We are chatting while passing several vegetable stalls run by migrants on the main street.

Shuhua: You always buy vegetables here?
Yang Cui: Yes, here vegetables are much cheaper! But not today . . . And here we can get real chilli.

Figure 8.1 The main street in the marketplace in Bomaqiao

Source: Photograph by Shuhua Chen

'Yes, here vegetables are much cheaper than there! But not today. . .' Yang Cui said. What Yang Cui meant by 'here' is that migrants, rather than locals, run the stalls. Yang Cui usually bought food from the migrant stallholders since, in general, they charged lower prices for their food than the locals did. Why, then, did she say, 'but not today'?

The 'but' comes from the concern about price and social face (*mianzi*). *Mianzi* literally means 'face' in Chinese. Face, as the first interactive layer of the body, is physical as well as social. The word '*mianzi*' is commonly used because of its social aspect – the face that can be gained, saved, enhanced, lost, or damaged through social interactions (Hwang 2012). The market is not only where people exchange food and capital but also serves as a place for people to meet and to interact during their daily shopping. Choosing to purchase food at the Bomaqiao market (instead of the supermarket) is not only a rational economic action but also involves great consideration of *mianzi*. It is different from the practice of food shopping in supermarkets, which is a choice mainly about the balance between one's food preference and the price.

Food shopping in the marketplace is tightly connected with family social status that may result in *you-mianzi* (having face) or *mei-mianzi* (having no face) (André 2013). *Mianzi* can be obtained through the way in which people perform a certain 'social code' that is publicly recognised by other villagers. In practice, a villager may feel *you-mianzi* (having face) when they buy expensive food from local stalls. By contrast, when a Bomaqiao local buys food from a migrant's stall, if the shopper bumps into villagers they know, they may feel they have no face. This feeling comes from the pressure or power of gossip among the villagers that shopping from migrants' stalls (which have food at cheaper prices) could be a sign that one cannot afford 'better' (in the sense of more expensive) local food. Thus, gossip around the village about one's family economic situation may start simply from buying food from a migrant's stall. Hence, food is not only a commodity in the marketplace but also a public signifier of one's social status.

Yang Cui's 'but not today' reflects this concern. In her understanding of how social face plays an important role in Bomaqiao local people's everyday lives, she was concerned about my social face, since, at that moment, to Yang Cui, I was a local because I was born in Shantou. The subtext is clear: 'I know that you (local people) care about your *mianzi* a lot'. As a migrant worker, Yang Cui believed that she did not know any local sellers, so there was no 'face' to maintain in her social interactions with them. In her everyday shopping by herself, she could buy food from whichever stalls she wanted – mainly migrant food stalls. However, on this occasion, she was shopping with me (a local), so she did not want me to be *mei-mianzi* (having no face) due to our buying cheaper food from the migrants' food stalls. Within such a fleeting moment of expression, her short statement 'but not today' provides us with a glimpse into how local and migrant perceptions of the same food can differ based on social and economic considerations. It helps us grasp the very process by which the food in the marketplace is shaped as a social and material phenomenon.

The tastes of home: 'That's my flavour of home!'

We then reach the junction of the locals' indoor market and the migrants' market street. We walk arm in arm, turning from the main street into the local market. Yang Cui points to a pork butcher stall, above which hang pieces of lean pork, belly slices, lard, and chitterlings. Neither the belly slices nor the lard interests me, but the fresh sea fish stall does.

Yang Cui:	I often buy lard here, cheaper than other stalls.
Shuhua:	Some fresh fish? I can cook for you today.
The fishmonger next to me:	I guarantee they are super fresh!
Yang Cui:	A fresh one? That should be expensive!

At the moment when the fishmonger responds to my question to Yang Cui, when we pass the fish stall, I find an expression of reluctance on Yang Cui's face. She is frowning and trying to find an excuse to 'escape' from the fish odour. A few steps away, a savoury smell wafts over from a braised goose delicatessen nearby that covers the fishy odour. We approach the deli counter, and we are lost in the various choices.

Shuhua:	That's my flavour of home!
Yang Cui:	Oh! Then let's go and buy some. How much?
The deli shopkeeper:	Meat 30 yuan one catty [one catty equals one half-kilogram], head 32, wings 35, feet 65. . .
Yang Cui:	I can't understand why the feet are so expensive!

Braised goose, a dish loved by most locals, is my favourite as well. Being a scholar of migration who has lived away from home (Shantou) for over twenty years and subsequently returned to carry out research, I have the personal benefit of being able to consider all kinds of food as 'comfort foods' (Abdullah 2010). During fieldwork, I usually ate with my co-workers, either meals at migrants' restaurants or meals cooked by Yang Cui. Unfortunately, even after a year, my stomach still could not adapt to the food that migrant workers enjoyed. They usually preferred food that I found too spicy and salty that always upset my stomach. I frequently experienced stomachache or diarrhoea after having an ordinary, everyday meal with them. Bomaqiao's local market, however, provided an especially intense sensorial environment as I encountered many of my favourite comfort foods, which revealed to me an intimacy with comfort food as my 'flavour of home' (ibid). From fresh *tegillarca granosa* (known as blood clam) to *oncomelania* (small tropical freshwater snails), various kinds of tropical coastal foods shape a unique foodscape.

Figure 8.2 A local butcher chopping pig feet

Source: Photograph by Shuhua Chen

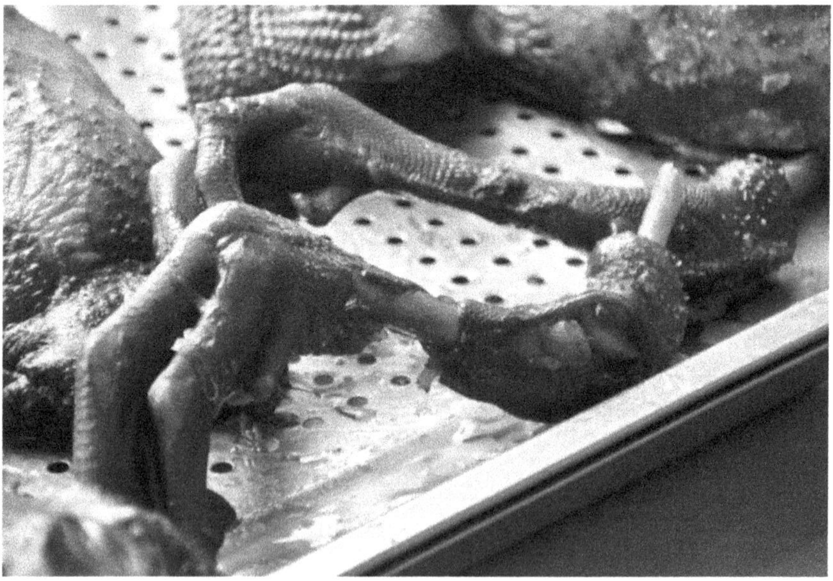

Figure 8.3 Expensive braised goose feet

Source: Photograph by Shuhua Chen

Figure 8.4 A local fishmonger picking fish for a customer

Source: Photograph by Shuhua Chen

Such a foodscape is so familiar to me that it creates a personal feeling of being at home.

However, most local foods that make my mouth water actually made Yang Cui feel sick: they reminded her of an earlier bout of diarrhoea she had experienced after having consumed local seafood. As for the pickled dishes, such as *gongcai*, the lack of any spicy flavour meant they were of very little interest to her.

Figure 8.5 Tegillarca Granosa, a side dish sold in the local market

Source: Photograph by Shuhua Chen

Figure 8.6 A shellfish stall in the local market

Source: Photograph by Shuhua Chen

Figure 8.7 A grocery shop selling *gongcai* (Chinese mustard pickles)

Source: Photograph by Shuhua Chen

Next to the grocery is a salt-baked chicken deli, which again makes my mouth water. I cannot help but try to buy some.

Shuhua: How about some salt-baked chicken?

Yang Cui: No spice at all. We can buy some fresh ones, and I will cook them for you, much cheaper and spicier!

We move on and stop in front of a fresh chicken stall. Yang Cui notices that I have been taking photographs while we are shopping. She asks me to take a photograph of the chickens lined up on the stall, similar to how quite often she suggested that I take notes on certain comments that she has made – or when she would warn me not to take notes.

Yang Cui: Come! Take [a photograph of] this one! You should take this one!

Shuhua: They are singing in silence, dancing in stillness.

I make a joke as a response to Yang Cui and try to persuade her again to take over the 'photography job' in my fieldwork. She refuses again, politely.

Figure 8.8 A chicken deli

Source: Photograph by Shuhua Chen

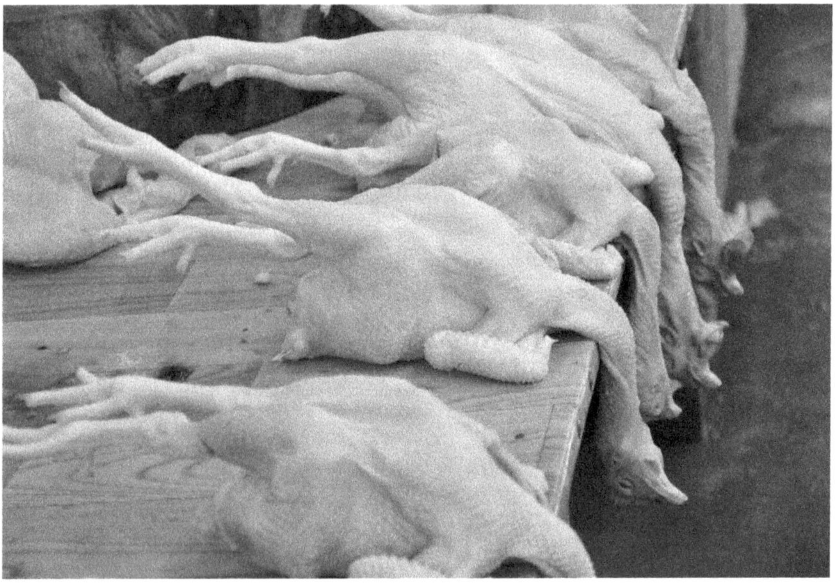

Figure 8.9 Chickens lined up on a local stand

Source: Photograph by Shuhua Chen

The practice of local food: 'No pig kidney again? No heart?'

While we are buying chicken, I hear a conversation at the next stall, which is a pork butcher.

A customer: No pig kidney again? No heart?
The butcher: It hasn't been available for these past two months. The Year of the Dragon brings us lots of dragon sons!

'No pig kidney again? No heart?' The customer asked with her face filled with surprise. Thinking that she had arrived at the market at such an early time (before 7:00 in the morning), maybe she assumed that she would have the chance to purchase some pig kidneys or hearts. I was surprised, too: Why are pig kidneys and hearts so popular in the local market? What has the Year of the Dragon to do with the scarcity of pig kidneys and hearts in Bomaqiao's food market? What are the particular foodways of pig kidneys and hearts in this particular context? Does the scarcity only hold true for kidneys and hearts from pigs, or does it also apply to organ meat from other animals?

The butcher explained to me how the story goes: in the Chaoshan region of South China (the broader cultural zone within which Shantou is situated), many local people believe in the food practice of *yixingzhixing*: eating a specific part of certain animals to strengthen the counterpart in the human body. For example, eating fish heads is believed to be effective in curing headache, eating pig kidneys is said to strengthen one's kidneys, and eating pig hearts nourishes one's heart. Part of their efficacy is thought to derive from their similarity in size to the corresponding human organs. So pig kidneys and hearts are very expensive in the local markets. Buying expensive food for mothers to be or young mothers is a way of showing one's love and great concern for them. The stock of pig kidneys and hearts available in the Bomaqiao local market could not meet the growing demand as there had been a baby boom in this year (2012), which was the Chinese Year of the Dragon. The dragons is regarded as a symbol of intelligence and strength in Chinese culture, and babies (especially boys) born in the Year of the Dragon are said to be destined to have a successful life.

Similar 'stories' could be easily found on the local market streets. Local beliefs and a traditional food therapy that emphasise 'the therapeutic effects of food, considering its nature, taste, and function on human balanced health' (Zou 2016: 1579) are the underlying foundation upon which the unique local foodscapes have been established in the Chaosan region. These foodscapes combined to create a bodily and sensorial kinship among its inhabitants, and the foci were always the market and home kitchen. The Bomaqiao market as the nexus of these broader social connections adapts to the changing tastes of its customers.

As we move on, Yang Cui and I pass a stall alongside the street, which is selling dog meat, heads, and organs, hanging on a simple steel pipe shelf. Next to the shelf is a signboard that states, 'Dog meat for sale'. Above the signboard, the seller lists detailed health benefits of each canine body part, such as:

> Eating dog head cures headache and dizziness
> Eating dog stomach cures gastric acid and stomach cold
> Eating dog testes and penis cures premature ejaculation and impotence
> Eating dog foetuses prevents miscarriage
> Eating dog meat boosts *yang* [in contrast with *yin*], enhances the *qi* [vital energy] balance, and strengthens bodies

After another few steps, there is a local alcohol shop selling various kinds of animal alcohol with different functions: namely, 'To relieve rigidity of muscles and expel wind-damp', 'To prolong life and anti-aging', and 'To nourish *yin* and strengthen *yang*'.

Seeing a local vegetable stall selling mustard leaf and Chinese broccoli, Yang Cui pulls me towards it:

Yang Cui: Some Chinese broccoli?
Shuhua: But aren't they much more expensive than over there?
Yang Cui: Yeah . . . But then will you be *mei-mianzi* [have no social face]? I know you people care for your [social] face a lot.
Shuhua: Oh, not me! Let's go buy the cheap one!

Figure 8.10 A local pork butcher stall

Source: Photograph by Shuhua Chen

Figure 8.11 The local market selling dog meat, heads, and organs

Source: Photograph by Shuhua Chen

Figure 8.12 A local alcohol stall

Source: Photograph by Shuhua Chen

Figure 8.13 Fish heads are expensive in the local market

Source: Photograph by Shuhua Chen

We smile at each other, understanding each other without any more words, and head to the next street – the migrants' market.

The experience of 'home' as a sensory totality within the migrant foodscape

Having spent less than five minutes in the migrants' market, Yang Cui has bought one Chinese cabbage, some spicy chillies, and some Sichuan *liangpi* (bean thread with Sichuan flavour). She shows great confidence and is very relaxed. She greets vendors here and there with her smiling face and enjoys tasting various 'home food' samples before buying them from food vendors on tricycles. She seems a different person than the one she was in the local market, where she was silent and felt uncomfortable. When we turn into the migrants' market street, she even volunteers to take over the role of the photographer. She tells me, 'I feel comfortable taking pictures here since we are not among locals'.

From the migrants' market, Yang Cui finds her 'home food' or comfort food quickly. It is perhaps similar to the way I can easily find my 'taste of home' from the covered locals' market. Since memories can be created and recreated in a sensorial milieu (Seremetakis 1994), the senses generated from the familiarity of foods on offer in the migrants' market may recall her original sensations as the

Figure 8.14 The migrants' market

Source: Photograph by Yang Cui

Figure 8.15 A vegetable stall in the migrants' market

Source: Photograph by Yang Cui

sense of being at home (Codesal 2010). Moreover, for Yang Cui, experiencing the migrant foodscape evokes her memory of home as a sensory totality. She enjoys the familiarity of the vendors' Sichuan accents when they greet her; she is attracted by the aroma of Sichuan *mala* (hot and spicy) flavour while enjoying tasting 'home food' samples offered by street food tricycles; she demonstrates

Figure 8.16 Yan Cui bought some spicy chillies from this stall

Source: Photograph by Yang Cui

Figure 8.17 Non-local street food sold by a migrant

Source: Photograph by Yang Cui

her consummate skill in selecting spicy chillies. Home is momentarily generated through her senses.

Sense of place and belonging: 'I feel comfortable to take pictures here'

For Yang Cui, the migrants' market is not only a place that sells familiar and comforting 'home food'; it is also a place where she feels at ease. What makes Yang Cui feel at ease there? From her narrative of her personal feelings ('I feel comfortable to take pictures here') to the reasons she has such feelings ('since we are not among locals'), the subject moves from an 'I' to a 'we': from an individual sense of 'I feel' to a collective identity of 'we are'. In her discourse, the 'we' not only refers to the migrant workers who come from the same village or the same province as Yang Cui; it also includes migrants from all over the nation, as long as they (like she) are not local. To some extent, the deictic 'we' (Hanks 1992) is a shared identity that is created to distinguish the non-locals (all the migrants) from the locals, the selves (we migrants) from the others (you locals). Such a separation is like an invisible wall, enclosing the migrants' market so that it is a space for the migrants – the non-locals – to have a sense of being at home, in terms of having a sense of the self and a sense of belonging. The differentiation of 'here' from 'there' through our discursive and physical movements functions to test and establish membership for the participants in spoken discourse within a shared framework of social reference (Schlegloff 1972). It is the 'we' (the migrants, the non-locals) who shape 'here' (the migrants' market) into a space of comfort and a place that is separate from 'there' (the locals' market), which is uncomfortable to her.

Our movement through the Bomaqiao market and, more importantly, our discussion of our movement, served to situate our discourse within a shared spatial frame. When I indicated my desire to buy the less-expensive Chinese broccoli, I was also indicating my wish to be included in the 'here' of Yang Cui's deictic reference system. By moving together through the space of the Bomaqiao market, we were *re-scaping* our sense of belonging: by purchasing food from the same shop, Yang Cui and I would come to embody a common if ephemeral identity through this shared process of mercantile and gastronomic consumption.

The ever-evolving texture of foodscapes

Before completing our shopping, Yang Cui makes a final stop to buy some fine dried noodles from a noodle shop run by migrants in the migrants' market. While local people regard rice as the staple food, for most of the migrants from other provinces, noodles are the staple of their diet. With more migrant workers moving to Bomaqiao in recent decades, the demand for noodles has increased, and more noodle shops have appeared in the migrants' market; noodles as a new staple are shaping the ever-evolving foodscape of the Bomaqiao marketplace. K. N. Chaudhuri's (1985: 25) observation about the 'dietary

Figure 8.18 A noodle shop run by migrants in the migrants' market

Source: Photograph by Shuhua Chen

punishment' experienced by early modern wheat eaters who were forced by their travels to eat other forms of carbohydrate would seem to apply equally here. Noodle shops are one of the few ways that wheat-eating migrant workers have to bring a comfort of home to this rice-centred coastal foodscape. As Mark Swislocki (2009) asks in his *Culinary Nostalgia: Regional Food Culture and the Urban Experience in Shanghai*, how does 'taste' register the change and continuity of a city's history? If we understand cities to be constantly changing 'migrant landscapes' (Rapport and Overing 2000: 379) that are transitory and diverse, then their fluid marketplace foodscapes play an essential role in crafting the ever-evolving texture of lives for locals as well as for the rural migrants in urban settings.

Urban migrant foodscape as a spontaneous form of tactics

It is 7:15am. We have done our daily shopping. It is time to head to the toy factory to start another twelve-hour workday. Like Yang Cui, many migrant workers in Bomaqiao also complete their daily shopping for food before beginning their morning shift. What is significant about their daily experience of the migrant foodscape? How might the everyday experience of the urban migrant foodscapes permit migrants to feel at home in cities? I would argue that for migrants, the foodscapes serve as a spontaneous enactment of what Michel de Certeau (1984) calls 'tactics' of everyday practices: a means by which to 'practice' home, even momentarily. For migrant labourers in Shantou, supposedly ordinary, everyday encounters within the urban foodscape take on extraordinary significance. Their lives are highly regimented: working hours run from 7:30am to 10:30pm, with only one day off per month. Labourers will frequently move from one factory to another in an effort to secure full-time work (Chen 2017). With so little time to themselves, there is little opportunity to make special arrangements or deviate from a daily routine. Living in cramped dormitories without privacy, the consumption of food becomes both a highly social and, as embodied experience, an intimately personal act. The urban foodscapes constructed and inhabited by the migrant communities of Shantou have become a powerful tactic of resistance – a means by which rural migrants savour a taste of home away from home.

Conclusion

To return the issue with which this chapter started, can a rural migrant ever feel at home in urban China? Through a tour of a marketplace, this chapter has shown that industrialisation and urbanisation of Shantou has 'scaped' the foodways of its inhabitants (locals, migrants, and myself, included as an anthropologist) in new and sometimes surprising ways that reveal a subtle sense of place, the constant negotiation of belonging, and vexed attachments to 'home'. I have argued that rural migrants are able to experience home, if only fleetingly, through the purchase of food in the marketplace. These purchases, and the social interactions that accompany them, permit them to enact a sensory 'scape' of home away from home. Though fleeting, the regular, quotidian nature of these interactions form a significant part of their lives away from home. Moreover, for migrants, the urban migrant foodscapes serve as a spontaneous form of tactics to 'practice' home in their everyday lives away from home. Constituted by the migrant vendors' shouts, the greetings exchanged between vendor and customer, the excitement of finding food from home, the choice of spices, the unique manner of food presentation, and the subtle sense of being at ease, to name just a few, the foodscapes provide a social space for the migrants to feel at home away from home in their everyday interactions in the marketplaces. It is manifest in the smells, feel, and taste of food from home as much as the sound of an accent from one's home province or the easy familiarity of shared experience. Home is, therefore, through its sensory totality, momentarily experienced.

References

Abdullah N. (2010) 'Comfort Food, Memory, and "Home": Senses in Transnational Contexts', in Kalekin-Fishman D. and Low K. (eds) *Everyday Life in Asia: Social Perspectives on the Senses*. Surrey: Ashgate: 157–176.

Ahmed S. (1999) 'Home and Away: Narratives of Migration and Estrangement', *International Journal of Cultural Studies* 2 (3): 329–347.

André J. (2013) 'How the Chinese Lost "Face"', *Journal of Pragmatics* 55: 68–85.

Chan W. (2010) 'The Global Financial Crisis and Migrant Workers in China: "There Is No Future as a Labourer; Returning to the Village Has No Meaning"', *International Journal of Urban and Regional Research* 34 (3): 659–677.

Chaudhuri K. N. (1985) *Trade and Civilisation in the Indian Ocean: An Economic History from the Rise of Islam to 1750*. Cambridge: Cambridge University Press.
Chen S. (2017) 'To My City: Urbanisation and Industrialisation in Contemporary China', *The Unfamiliar: An Anthropological Journal* 7 (1): 27–29.
Chen S. (2018) 'Homeawayness and Life-Project Building: Making Home Among Rural-Urban Migrants in China', in Nicola F. and Selwyn T. (eds) *Travelling Towards Home: Mobilities and Homemaking*. Oxford: Berghahn Books: 34–54.
Codesal D. (2010) 'Eating Abroad, Remembering (at) Home: Three Foodscapes of Ecuadorian Migration in New York, London, and Santander', *Anthropology of Food* 7: NP.
De Certeau M. (1984) *The Practice of Everyday Life*. Berkeley: University of California Press.
Hanks W. F. (1992) 'The Indexical Ground of Deictic Reference', in Alessandro D. and Goodwin C. (eds) *Rethinking Context: Language as an Interactive Phenomenon*. Cambridge: Cambridge University Press: 43–76.
Hwang K. (2012) 'Face and Morality in Confucian Society', in *Foundations of Chinese Psychology, International and Cultural Psychology*, Vol. 1. New York: Springer.
Ingold T. (2011) *Being Alive: Essays on Movement, Knowledge and Description*. New York: Routledge.
NBSC. (2019) 'National Bureau of Statistics of China 2018 Report of Migrant Workers in China', www.stats.gov.cn/tjsj/zxfb/201904/t20190429_1662268.html.
Olwig K. (2005) 'Representation and Alienation in the Political Land-Scape', *Cultural Geographies* 12 (1): 19–40.
Rapport N. and Overing J. (2000) *Social and Cultural Anthropology: The Key Concepts*. Abingdon: Routledge.
Seremetakis C. N. (ed) (1994) *The Senses Still: Perception and Memory as Material Culture in Modernity*. Chicago: University of Chicago Press.
Schlegloff E. (1972) 'Notes on a Conversational Practice: Formulating Place', in Giglioli P. (ed) *Language and Social Context: Selected Readings*. Baltimore: Penguin Books: 95–135.
Stoller P. (1989) *The Taste of Ethnographic Things: The Senses in Anthropology*. Philadelphia: University of Pennsylvania Press.
Stoller P. (1997) *Sensuous Scholarship*. Philadelphia: University of Pennsylvania Press.
Swislocki M. (2009) *Culinary Nostalgia: Regional Food Culture and the Urban Experience in Shanghai*. Stanford, CA: Stanford University Press.
Zou P. (2016) 'Traditional Chinese Medicine, Food Therapy, and Hypertension Control: A Narrative Review of Chinese Literature', *The American Journal of Chinese Medicine* 44 (8): 1579–1594.

9 Sourcing, sensing, and sharing Bengali cuisine on the Gold Coast

Diti Bhattacharya

There is something distinct about the smell of boiling *Govinda bhog chaal* (a variant of rice grain produced primarily in the Damodar basin of West Bengal, India). The pungent aromatic steam is significantly different from that of other rice variants from the region. Its grains are small and rounded, becoming pulpy, sticky, and starchy when cooked. It is this distinct pulpy texture which made it a regular of the everyday Bengali kitchen – it was homely, intimate, and familiar. Among the many things I was going to miss migrating to Australia, *Govinda bhog chaal* was on top of the list. Months later, my joy knew no bounds when I noticed my roommate on the Gold Coast cooking the same sticky, pulpy rice. What I found confusing was that Samena referred to *Govinda bhog* rice as *Kalijira* rice from Dhaka. My staple lunch in Kolkata had a namesake in Bangladesh.

Samena and I were international students at Griffith University in Australia between 2012 and 2015. Born into a Muslim Bengali family, Samena is from Dhaka, the present capital of Bangladesh. In contrast, I am from Kolkata (previously Calcutta) in West Bengal, India, born into a Hindu Bengali family. Our paths crossed on a windy winter afternoon at a student accommodation in Labrador, a quiet suburb in northern Gold Coast, south of Brisbane, Queensland. Of the five university campuses, we were quite predictably attracted to the Gold Coast's pristine white beaches and its shiny skyscraper buildings. We shared an apartment as international students for a year. Oscillating between being tourists and locals, we explored the Gold Coast while also making sense of it as our new home. We would try new cuisines at restaurants, and we bonded in conversations over the food we had left behind – home-cooked Bengali food – which became the foundation of our budding friendship. Gradually we began cooking together, and we realised that home is not what it looks like, but what it smells, tastes, and feels like.

This chapter explores sourcing, sensing, and sharing raw ingredients for ethnic Bengali dishes as an affective, sensorial, and material process. Drawing on the differences and similarities of our individual and collective memories as well as our cultural and geopolitical subjectivities, I examine how our culinary memories and knowledge of two different homes (Dhaka and Kolkata) recreated a new collective sense of belonging at the Gold Coast in Australia. This chapter comprises two parts. First, reflecting on the differences and commonalities that emerge from our culinary practices, I map the ways in which our lived diasporic experiences are mediated through gastronomic materialities. I suggest this process can dissolve emotional barriers of geopolitical complexities and socio-cultural stereotypes in creating a collective sense of home. Second, I argue that this process of creating a food-centred 'sense of home' aids in rethinking ways in which memory and city spaces negotiate their relationship.

Existing literature at the intersection of human geography, feminist geography, and geographies of food has continually explored the complex human, more-than-human, and sensorial entanglements between everyday lived experiences of migrants and gastronomic materialities (Anderson and Smith 2001; Ashley *et al.* 2004; Goodman 2016; Holt 2017; Janeja 2010; Longhurst *et al.* 2008). While Robyn Longhurst, Elsie Ho, and Lynda Johnston have examined culinary practice as a 'political issue' that aids in 'resituating and reconstituting diasporic subjects' (2008: 333), Michael K. Goodman stresses the importance of analysing 'affect, embodiment and cultural practices' in understanding the 'multiple materialisms of food, space and eating' (2016: 257). In this chapter, I take David Sutton's gustemological approach by paying attention 'to the sensory aspects of eating, not just in the moment of now, but in the deeply evocative ways that food can tie together multiple strands of the past that infuse any present social situation' (2010: 473–474). Such an approach 'offers new ways of thinking both about selfhood and subjectivity and about group identity and collective memory' (ibid). The chapter will contribute to the emerging and existing conversations between individual and collective memories and materialities of sensory experiences and foodways. It will do so by rethinking how embodied methods reveal food politics as part of migrant subjectivities' references to and making of home.

Ghassan Hage defines home building as 'the building of the feeling of being *at home*' (2010: 417). This process of 'building of the feeling' is informed by an intimate and careful choice of memories of the past, threaded together with the experiences of the present. The process of threading together these memories and experiences is not limited to the kitchen but often spills over to other public spaces of the city. For Samena and myself, our conversations and revelations around sharing the same culinary culture and practices were never contained within the kitchen only – our stories from the past often emerged unexpectedly on our walks to the grocery stores or in between trying some new food in a restaurant. This chapter's focus is on this coming together of our diasporic memories of food through a sensorial and embodied experience of city spaces.

Methodology

My research is grounded in thinking about the ways in which our bodies can be used as research instruments in understanding the relationship between food, spaces, and memories. As I explore a 'culturally embodied difference' (Longhurst *et al.* 2008: 209) through cooking and eating, my ethnographic reflections are primarily focused on the ways in which bodies undergo a material and sensuous experience while being in the field. A recent interest in the geographies of emotion has led to a renewed focus on how bodies are the primary point of intersecting emotions and spaces (Anderson and Smith 2001; Laurier and Parr 2000). Remembering our experiences of certain food items that we consumed individually in Dhaka and Kolkata and then recreating those moments in the present unfolded through our bodies. As John-David Dewsbury and Simon Naylor contend, the field can never be fully explored 'without the performance of bodies and materialities to define its boundaries' (2002: 256). The body remembers complex and messy entanglements of political events, city spaces, and everyday food practices together as an assemblage. A major aspect of my methodological tactic was to conjure these complex embodied memories through a 'sensuous ethnography' (Stoller 2004). My ethnographic reflections are often nonlinear, dynamic, and messy because they prioritise the emotions of our culinary practices as remembered through our embodied experience. The chapter comprises three sections. Firstly, I provide a brief geopolitical context of being Bengalis from Dhaka and Kolkata. Secondly, I map the everyday material and sensorial and embodied experiences that we encountered as Bengali migrants on the Gold Coast through sourcing ingredients and cooking. Finally, I explore the ways our present culinary practices and memories from two different cities reshape and reproduce our own sense of home on the Gold Coast.

The historical context of Bengaliness

Samena's and my Bengaliness is characterised by complex and differing geopolitical, cultural, and affective layers. This difference is based on our places of origin – Dhaka and Kolkata – which influenced our self-perception as Bengalis and our culinary practices. Through our process of 'homemaking' on the Gold Coast, we engaged with our memories of Dhaka and Kolkata through food. We realised that the sensorial registers food, and city spaces create a 'presently existing space of the past' (Hage 2010: 427). This space of the past that we continually yearned to create in the present is informed by political and historical memories as well as their socio-cultural impacts on culinary practices. While a thorough examination of the colonial and communal past of India and Bangladesh is beyond the scope of this chapter, in the following paragraph, I briefly trace the political events that affect the complex identities of the two cities and their culinary practices.

One of the most significant consequences of the colonial violence was the division of Bengal – a state in the Eastern part of India and the British capital of the country for two hundred years. Before 1905, contemporary Bangladesh and West Bengal in India were one single state within India during British colonial rule. Bengal's first territorial division came into effect in 1905 under the leadership of Lord Curzon, in which a Hindu majority West Bengal and a Muslim majority East Bengal (now Bangladesh) came into existence still as part of the Indian territorial region. While the people living in both the sectors identified themselves as Bengalis, the politics of being a Bengali began to complicate itself. Following political instrumentalisation of Bengalis as an ethnic group, the communal riots of 1946 witnessed an influx of a large number of Hindu and Muslim Bengalis from East Bengal to West Bengal. This period of forced migration fuelled by sectarian violence altered the demographics of West Bengal into a Hindu majority state. At the same time, East Bengal became a Muslim-Bengali majority state. The Bengalis of East Bengal later came to be known as *batis* or *bangal*, while the existing Bengalis of West Bengal identified themselves as *ghotis*. In 1947, India became independent from British colonial rule, a process that went hand in hand with Indian partition on religious grounds. While West Bengal remained part of India, the new state, Pakistan, comprised two separate geographical territories: West Pakistan (present-day Pakistan) and East Pakistan (contemporary Bangladesh). Even though the first prime minister of India, Jawaharlal Nehru, 'expressed strong opposition to attempts to politicise the country's ethnic, religious and language diversity' (Shamshad 2017: 434), the socio-political fabric of Bengal and India had long changed as a consequence of the continuous colonial (both East India Company and later the British Raj) policies, which encouraged politicisation of ethnic differences. These divisions remain entrenched and are still a part of our everyday socio-cultural narratives, including food narratives.

In 1971, two wars raged on the Indian subcontinent: a civil war between East and West Pakistan and a war fought between West Pakistan and India. In both, wars 'ethnicity colluded with national interests and state politics' (Saikia 2014: 275). The Urdu-speaking Muslim community (primarily Muslim migrants from the state of Bihar, India, who moved to East Pakistan in 1947), the Muslim League, and a noticeable number of Muslim Bengalis supported West Pakistan. These were defeated by the *Mukti Bahini* (Liberation Army) with the support of the Indian army, which 'resulted in the partitioning of Pakistan and creation of an independent nation-state of Bangladesh' in the territory of East Pakistan (Saikia 2014: 274).

Evidently, these political events left their mark on the everyday socio-cultural textures of cities. These differences and commonalities in our respective Bangladeshi and Indian Bengaliness trickled down into Samena's and my understandings of Bengali food. Our individual and collective understanding of Bengali culinary practices became more complex through our experience of migration. While it is easy to perceive us as Bengali migrants in Australia, as a unified entity, our political and cultural existences are distinct and multiple overseas. A simple search on the internet shows that the Bangladeshi Association of Queensland (comprising migrants from present-day Bangladesh) separates itself quite distinctly from the Bengali Society of Queensland (consisting of migrants from West Bengal, India). This distinction is not only prevalent in community-based organisations but is also mirrored in everyday practices. The tension between two different kinds of Bengalis is a socio-political consequence of the colonial rule experienced by India. Over time, this has spilled into the gastronomic traditions of both Bangladeshi and Indian (such as Bengalis from West Bengal) migrant communities. Several Bangladeshi restaurants and grocery shops are distinctly different from their Indian counterparts in terms of the items they serve and sell. The Bengalis of Bangladesh use certain elements and implement cooking styles different to those of Bengalis from West Bengal. During our initial days of getting to know each other, Samena and I struggled to negotiate these complexities – being 'Bengali' was messy for us.

Despite these differences, there are also important commonalities. Both Bengali subjectivities share 'a collective proper name, Bengali or *Bangali*; shared historical memories; an association with a specific homeland, United Bengal/Bangla/Banga; and a common language, Bangla or Bengali' (Shamshad 2017: 435). Similarities overlap into everyday culinary practices, in which, for example, Bengalis eat the same staple food even though different parts of Bengal practise different cooking styles and pursue individual religious inclinations despite living in the same state.

Even though ethnic and religious clashes fuelled the division of Bengal, they ended up carving a cultural and political legacy that prioritises a sense of 'being Bengali' beyond religious, ethnic, and political difference (Mookherjee 2008). Both Hindu and Islamic traits influence Bengali music, art, culture, and food practices. While Bangladesh was considered to be an Islam majority nation, the Islam of Bengali Muslims was starkly different from the Islam of West Pakistan, in terms of its religious essence, culture, and practice (Mookherjee 2008: 60). This also had significant culinary influence in which, for the *Bangals* (migrants from East Pakistan), cooking styles and sporting activities became a central place of holding on to their Bengaliness (Bandyopadhyay 2008). At times, similarities became more meaningful than differences while alternatively, differences helped migrant Bengalis assert their history amidst other Bengalis who had never experienced forced displacement.

Cultural exchange between communities is restrictive with respect to social and cultural events. Samena has never attended the Bengali events that I participated in with the Bengali (from West Bengal) diasporic community, nor have I participated in activities of the Bangladeshi community. There has always been a hesitation for these events or get-togethers. Yet through our shared culinary practices on the Gold Coast, we noticed how gastronomic materialities and their affective and sensorial power facilitates a blurring of lines – a renewed sense of enthusiasm in knowing each other slightly better, in contrast to hesitation. Arjun Appadurai notes, in his discussion of gastro-politics in Hindu South Asia, that a potent characteristic of food is 'its capacity to mobilise strong emotions' (1981: 494). Food not only mediates emotions but also mobilises these emotions towards fostering a more profound sense of affective material attachment in creating a sense of feeling at home. Similarly, Allison Hayes-Conroy and Jessica Hayes-Conroy use the practice of eating and food to redefine what they call 'visceral geographies' (2010: 334). In our Gold Coast kitchen, there was not only a coming together of ingredients through cooking but also an assembling of two different cities. The process was multi-sensorial and layered because as we cooked, we shared stories of Dhaka and Kolkata that we had heard from our family members. One can argue that these discussions are possible anywhere. Is it merely coincidental then that we exchanged our stories while cooking? I suggest otherwise. Food is unique in how it can ignite specific sensorial registers. Through the process of creating stories in domestic cooking spaces, such as the kitchen or shared public spaces associated with culinary engagements such as grocery shops, restaurants, or farmers' markets, our food discussions provided the bridge for making spaces familiar in the present. Thus, feelings of being at home were not restricted to the kitchen or dining table but were also created in shops, spice lanes, and on bus journeys to these places.

Sourcing ingredients: braiding in difference and familiarity

Govinda bhog chaal is harvested in the Burdwan and Bankura district of West Bengal. In contrast, *Kalijira chaal* grows best in the Dinajpur district of Bangladesh. Samena and I often prepared this variant of rice: on some days, we cooked *Govinda bhog*, and on others, we cooked *Kalijira* rice. We noticed how the raw versions of the two variants looked the same but produced slightly different flavours when cooked. Revisiting our individual memories of taste and smell of dishes that complemented the rice, we paired our version with either *biuli r daal* (black gram lentils) or *phalon daal* (a type of lentil found only in Bangladesh), *aloo posto* (potato and poppy seed curry) or *shutki bhorta* (dry fish curry).

While our recipes were simple, the preparations were elaborate. We made the effort to travel to a particular grocery shop in Inala – about a 40-minute bus ride from Brisbane City – to source the right seeds for the *daal* (lentils) or the copper utensils. The ways in which the individual shop rows were organised, including the congested piling of spice packets next to the pile of incense packets, reminded us of the roadside grocery stores in Dhaka and Kolkata. Our culinary practices were laced with numerous stories of our mothers and grandmothers and how they taught us to cook. Stories that reminded us how things were done at home informed our senses and closely monitored how we shopped, cooked, and ate together. Samena would often recount how her mother never threw away rice starch as this imparted a thick note of flavour to the cooked rice, reminding me of my grandmother's cooking style. We made sure that we heated fenugreek seeds to the right temperature so they would impart a distinct fragrance to the boiled *biuli r daal* seeds when cooked together. The smell of the *panch phoron* (Bengali five-spice) heating up, the starchy rice bubbling in the pot, coupled with our giggles oscillated between an imaginary of 'feeling at home' and a present of creating home.

Sourcing fresh-river fish varieties in and around the Gold Coast was always a struggle. The closest we could get was buying dried varieties of Bengali fish, such as *loita shutki* or *chingri shutki*, which were traditionally popular among the Bangladeshi community. We could never find these varieties in the Indian grocery shops, but the Bangladeshi grocery shops in Logan City (an hour-long bus journey from the Gold Coast) always had them. Samena often teased me about how the Bengalis of West Bengal frowned on the consumption of dried fish and only preferred freshwater varieties. I could not deny this and remembered how my family members often complained about the strong, pungent smell of cooked dried fish curries. When on some afternoons, the smell of dried fish curry came through the kitchen windows of a neighbouring house, my grandmother often wondered if a *Bangal* had moved into our neighbourhood. I could never determine if her tone connoted a mere curiosity or a sense of disapproval. As much as I tried to come up with a counterargument, here in Australia, where there was a shortage of fresh Rohu fish, I secretly craved the dried fish curry that Samena would make. After all, it was still Bengali food. This was the first instance I became hopeful of re-uniting and reliving my memories using the tactile process of cooking and sharing food through a food item that was very much Bengali but also unfamiliar to me. Samena and I would often muse over creating similar environments in our 'home' kitchen here on the Gold Coast. However, we never quite knew what that fleeting moment would look like.

Cooking would move our senses in ways that were both known and unknown to us. We both adapted to tastes and smells that belonged to Dhaka and Kolkata. The process of adaptability in diasporic transition and settlement experience is important here. Describing a Lebanese woman's joy and excitement at being able to purchase Lebanese cucumbers grown in Sydney, Hage explains 'how the practices of fostering intimations of being in Lebanon (represented by making a salad with cucumbers to yield their potential homeliness) are at the same time practices of home-building in the here and now' (2010: 424). The differences and similarities that were emergent through our process of sourcing ingredients were moments of hopefulness grounded in sensing urban space and the food culture we had left behind, to recreate that sense through our own present, shared experiences. On numerous occasions, we failed to find authentic ingredients. At other times, I could find ingredients familiar to myself that Samena was not accustomed to, and vice versa. However, it gradually became evident that the yearning to produce a 'sense of home' through food was not about replicating recipes but instead about emotionally choosing and settling with alternatives to feel a sense of togetherness in the present. These feelings were not always restricted to the kitchen of our apartment but were spontaneous and often occurred while we looked for grocery shops on Google Maps, during short conversations with an Indian or Bangladeshi shopkeeper, or when we chanced upon a known food item while shopping or brainstorming what replacement ingredients we would use to create a particular dish.

Our embodied memories of food were often recreated through walking and exchanging notes on how we remembered the sensorial aspects of raw ingredients back in our kitchens in Dhaka and Kolkata. Walking through the spice lanes of the grocery shops, the smell of the raw *panch phoron* triggered images of hot oil in a wok in our imaginaries. On another occasion, Samena spent hours trying to identify the right kind of red chilli from the area that she belonged to in Bangladesh. She described the distinct vibrant red colour of the chilli paste that imparted the pungent smell of chilli and red gravy of a fish curry cooked often in her house. We anticipated and guessed how we would be able to bring in the taste here; we worried and joked about the fire alarm going off in the kitchen. What became more important was the process of assembling, trying, testing, and rediscovering memories of taste, sense, homeliness, comfort, and security through food in a new city space. Ben Anderson and Colin McFarlane note that 'to understand assembling as a process of "co-functioning" whereby heterogeneous elements come together in a non-homogeneous grouping' (2011: 125). Feelings of 'being at home' through food are not always about replicating that which is in the past. It is a process of combining and assembling raw ingredients, cooked food – and through the memories that emerge being moved viscerally. Discussing visceral geographies of food, Hayes-Conroy and Hayes-Conroy note that 'eating – due to its sensual, visceral nature – is a strategic place from which to begin to understand identity, difference and power' (2008: 462). The moments we created while shopping for ingredients, cooking together, and eating food were not only instrumental in making us feel at home but also mediated our new identities as complex assembled Bengali migrants. In this context, using our bodies to remember the raw ingredients became crucial. In a different country, we adapted to a newer set of sensory and material registers. However, our embodied memories of 'home' always remained resilient and deeply etched in our memories. These moments of 'remembering' the sensorial registers of the past are often aids in the processes of adaptability, adjustment, and recreation of home among migrants, like us.

In our collective experiences over time, it became evident that being Bengalis together was to create a sense of home that evoked the pungent smell of garlic, chilli, ginger, and cloves, typical of both our kitchens back in our respective 'homes'. We wanted to bring together and create an assemblage of smell, colour, and taste that was familiar to both of us. For example, when Samena cooked dried fish curry with strong Bangladeshi *batasha longka* (a kind of chilli grown in the Khulna region of Bangladesh), the smell of combining ingredients, with which I was not very familiar, did not fail to remind me of home. Despite our Bengaliness – including differences and similarities – negotiating these brought together 'new food desires and attachments' (Hayes-Conroy and Hayes-Conroy 2010: 1280). Through this process, we also engaged in re-registering sensorially our individual kitchen spaces back in Dhaka and Kolkata. For us, the kitchen smelling like the 'home kitchen' was as important as the food tasting like home-cooked food. Gradually, over time, we knew we belonged together as the individual ingredients and various cooking styles brought together an assemblage that created strands of familiarity. These emotions of belonging do not merely reside in subjects, historical events, or geographical borders in isolation. Moving through visceral acts of cooking together, eating together, and in between bodies (Johnston and Longhurst 2012: 328), they indicate a sense of ease in the formation of new cultural threads within diasporic contexts.

While the cooking of dry *loita* fish (*loita shutki*) and dry prawn fish (*chingri shutki*) originates in the Chittagong area of Bangladesh, the *Rohu Kalia* (freshwater Ruhu fish curry) is more characteristic of the *ghotis*. Even though both recipes were similar in the sense that they were semi-dried curries slow-cooked with ginger, garlic, tomato, and garam masala, the dried fish variety hardly ever made it to the kitchen of the Kolkata Bengalis. My grandmother often commented on the dry fish cooking practices of the *Bangals*, associating their financial and social struggles with their inability to afford fresh fish regularly: hence, resorting to dried fish. Years later, sharing my life as a 'new' migrant with Samena, I realised how my privileges of being a Bengali from West Bengal, someone who had never experienced migration until now, haunted me. There was a sense of guilt that arose from realising my privilege of never having to experience forced migration, unlike the *Bangals* in West Bengal. Yet this guilt was accompanied by a curiosity to try a different style of fish cooking which forms part of Bengali cuisine.

Bringing together the sensory and culinary: between Kolkata, Dhaka, and the Gold Coast

Samena often pointed out that the adaptive quality of cooking multiple versions of the same item came naturally to people who have experienced migration and displacement. In contrast, for me, coming from a *ghoti* Bengali family, retaining the authenticity of how a particular dish must be cooked was more critical. This perception changed with my own experience of migration on the Gold Coast. A seemingly unknown city began altering my motivation for using a number of ingredients. My experiences and, at times, struggles with creating a new home while making sense of my past had made me realise that 'authenticity' is essentially a cultural construct laced with complex power relations (Lindholm 2002). Being authentic is often associated with being 'sincere, true, honest, absolute, basic, essential, genuine, ideal, natural, original, perfect, pure'. (ibid: 363). Born into a family that was originally from West Bengal, having no history of migration before I moved to Australia, I was very familiar with how 'our' style of choosing ingredients and cooking was different from 'theirs'. These differences are also reflected in a language of othering through food in everyday *ghoti* colloquial language. For me, this sensory distinction also shaped my palate. For example, the *ghotis* like myself are known to put copious amounts of palm sugar in their food. A *ghoti* family's staple lunch is rice; *daal* (lentil soup) made out of black Bengal gram seeds; and *aloo posto*, a curry made out of white poppy seeds and potatoes. The *ghotis* believe that it is possible to achieve a sense of class-based hierarchy by cooking food that induces a sense of comfort, slowing down, and sweetness.

In contrast, after the influx of migrants from East Bengal during the 1940s and in 1971, during Bangladesh's liberation movement, the markets in Calcutta began to stock dried fish, taro leaves and stems, red spinach, and giant river catfish, among other items unique to the Eastern side of Bengal. The *Bangals* found refuge both gastronomically and metaphorically in raw materials and a cooking style that was more 'hard-earned' and 'sustainable'. For example, *kochu diye kanchkolar torkari* (taro and green banana curry) – a common dish among the migrant Bengali communities – is primarily known as a curry that uses every element of the ingredients in the cooking process. The green banana and taro are cooked in cumin, coriander, ginger, and turmeric, along with the skin, which is used as a crispy textured layer in serving the curry. For the *Bangals* who were trying to make sense of their suddenly acquired refugee status in West Bengal, at different points of time, maintaining and practising these culinary differences were integral to maintaining their Bengaliness.

These food-centred tensions that we grew up with in different cities never left us. However, our individual experiences of migration aided us in rethinking what we could do with these stories. Food and its allied culinary practices were a sensorial and affective vessel to remember our culinary past and to create a sense of home in the present. The thing about food that is significant in this context is that it is both tactile and mobile. The anticipation and excitement we experienced in bus rides to grocery stores in and around the Gold Coast were mobile sensorial moments of a certain kind of coming together of what we had experienced in Dhaka or Kolkata and how that experience might re-unfold in a new city.

Within the context of our experience, this process of 'tying together' of numerous strands has been not just about food, but also about cities. The memories of food that we carried forward and shared through individual ingredients or cooked food items are in many ways exalted memories of the cities we lived in – Dhaka and Kolkata – to bring into the city that we were trying to make our own, the Gold Coast. 'Context' as Nigel Thrift explains is 'not . . . an impassive backdrop to situated human activity', rather it is 'a necessary constitutive element of interaction, something active, differentially extensive and able to problematise and work on the bounds of subjectivity' (1996: 3). In the process of creating stories in domestic cooking spaces, such as the kitchen and the kitchen counter, or in shared public spaces associated with culinary engagements, such as grocery shops, restaurants, and food markets, our food stories and discussions provided the 'context' for making these spaces familiar to us in the present. Thus, feelings of being at home were not just restricted to the kitchen or the dining table but were also experienced at the grocery shops, in the spice lanes, and on the bus journeys within the city of the Gold Coast. As we moved in search of food ingredients or utensils, we encountered memories through smell, touch, taste, and colour of our 'homes' back in Dhaka and Kolkata. In this sense, a complex nexus is formed between culinary practices, the senses, and three different cities that were all part of us simultaneously. Samena began to understand what Bengali food in Kolkata meant, whereas I learned about Bengali food in Dhaka – while both of us were living and eating together on the Gold Coast. This realisation came through our process of creating new narratives by weaving and reweaving in the old ones.

Conclusion

In this chapter, I have considered affective sensorial and material entanglements through the process of sourcing ingredients and cooking. I have traced individual and collective experiences and memories to show some ways to rethink homemaking. The chapter offers an analogy for how interactions between place, materiality, and gastronomic practices occur in transitionary sites such as the kitchen, streets, or grocery shops. It demonstrates how assembled experiences blur geopolitical complexities and socio-cultural barriers, especially in subjectivities with migration experience, such as Samena and I. For us, it is the kitchen; the raw ingredients; and the processes of remembering, forgetting, re-remembering, and recreating our food histories and practices that contributed to intense emotional homemaking. Finally, I have explored how homemaking on the Gold Coast became a multi-layered process for us – oscillating between the homes we left behind in Dhaka and Kolkata and our present home on the Gold Coast through our discussion of, shopping for, cooking, and consuming our culturally laden cuisines. This chapter suggested that embodied processes of negotiating culinary memories help to redefine people's relationships with city spaces through a sensorial aesthetic.

References

Anderson B. and McFarlane C. (2011) 'Assemblage and Geography', *Area* 43 (2): 124–127.

Anderson K. and Smith S. (2001) 'Editorial. Emotional Geographies', *Transactions of the Institute of British Geographers* 26: 7–10.

Appadurai A. (1981) 'Gastro-Politics in Hindu-South Asia', *American Ethnologist* 30 (1): 494–511.

Ashley B., Hallows J., Jones S. and Taylor B. (2004) *Food and Cultural Studies*. London: Routledge.

Bandyopadhyay K. (2008) 'The Nation and Its Fragments: Football and Community in India', *Soccer and Society* 9 (3): 377–393.

Dewsbury, J.-D. and Naylor S. (2002) 'Practising Geographical Knowledge: Fields, Bodies and Dissemination', *Area* 34 (3): 253–260.

Goodman M. K. (2016) 'Food Geographies I: Relational Foodscapes and the Busy-Ness of Being More-Than-Food', *Progress in Human Geography* 40 (2): 257–266.

Hage G. (2010) 'Migration, Food, Memory and Home-Building', in Radstone S. and Schwarz B. (eds) *Memory: Histories, Theories, Debates*. New York: Fordham University Press: 416–427.

Hayes-Conroy A. and Hayes-Conroy J. (2008) 'Taking Back Taste: Feminism, Food and Visceral Politics', *Gender, Place and Culture* 15 (5): 461–473.

Hayes-Conroy J. and Hayes-Conroy A. (2010) 'Visceral Geographies: Mattering, Relating, and Defying', *Geography Compass* 4 (9): 1273–1283.

Holt L. (2017) 'Food, Feeding and the Material Everyday Geographies of Infants: Possibilities and Potentials', *Social & Cultural Geography* 18 (4): 487–504.

Janeja M. K. (2010) *Transactions in Taste: The Collaborative Lives of Everyday Bengali Food*. Abingdon: Routledge.

Johnston L. and Longhurst R. (2012) 'Embodied Geographies of Food, Belonging and Hope in Multicultural Hamilton, Aotearoa New Zealand', *Geoforum* 43: 325–331.

Laurier E. and Parr H. (2000) 'Disability, Geography and Ethics: Emotions and Interviewing in Health and Disability Research', *Philosophy and Geography* 3 (1): 98–102.

Lindholm C. (2002) 'Authenticity, Anthropology, and the Sacred', *Anthropological Quarterly* 75 (2): 331–338.

Longhurst R., Ho E. and Johnston L. (2008) 'Using "the Body" as an "Instrument of Research": Kimch'i and Pavlova', *Area* 40 (2): 208–217.

Mookherjee N. (2008) 'Culinary Boundaries and the Making of Place in Bangladesh', *South Asia: Journal of South Asian Studies* 31 (1): 56–75.

Saikia Y. (2014) 'Ayub Khan and Modern Islam: Transforming Citizens and the Nation in Pakistan', *South Asia: Journal of South Asian Studies* 37 (2): 292–305.

Shamshad R. (2017) 'Bengaliness Hindu Nationalism and Bangladeshi Migrants in West Bengal, India', *Asian Ethnicity* 18 (4): 433–451.

Stoller P. (2004) 'Sensuous Ethnography, African Persuasions, and Social Knowledge', *Qualitative Inquiry* 10 (6): 817–835.

Sutton D. E. (2010) 'Food and the Senses', *Annual Review of Anthropology* 39: 209–223.

Thrift N. (1996) *Spatial Formations.* London: Sage Publications.

10 Transmitting traditions

Digital food haunts of Nepalis in the UK

Premila van Ommen

Digital sensory encounters of food via social media are inescapable for many British Nepalis. These frequent encounters reveal patterns that articulate how a diaspora perceives themselves and reflect ways they mark space in the city. This chapter examines digital practices of Nepalis in Britain to uncover relationships and understandings between urban, migrant place-making and identity formation through the medium of food. Based on ethnographic methods and textual analysis of social media content from Nepalis in London aged between 20 and 35 since 2014, my investigation includes auto-ethnography and online participatory research. This chapter argues that ephemeral media and other evolving technologies in social media applications contribute new dimensions to ethnographic approaches and analysis that include better understandings of sensorial food encounters, embodied practices, and conviviality. Temporal aspects of digital media reveal routes where senses of space and distance collapse through place-making presented by images of physical diasporic food sites. The ephemeral alongside permanent content uncovers variances between intentions and interpretations of uploads surrounding food topics and places. Layering meanings into a diasporic consciousness built out of Gurkha military migration histories, these contents flow along digital streams to facilitate multiple senses of what it means to be Nepali.

This research explores food's roles in mediatised discourses informed by Mirca Madianou and Daniel Miller's (2012a, 2012b) developing theory of polymedia. Polymedia is 'an emerging environment of communicative opportunities that functions as an "integrated structure" within which each individual medium is defined in relational terms in the context of all other media' (Madianou and Miller 2012b: 170). Polymedia theory takes into account the multiple ways individuals shift between forms of technology according to strategies based on degrees of emotional investment. Instead of focusing on specific technologies and their affordances, polymedia focuses on networks and systems of media usage. By first outlining the demographic makeup of Nepalis in Britain, this chapter sets their polymedia environment within a background of indigenous identity politics and colonial military history. This chapter then unfolds narratives within British Nepali creations and consumption of digital food-related content. It interrogates how they feed networks connecting Nepali food places and chart developments in diasporic place-making, adding sensory dimensions to the city from London to Aldershot by drawing new maps from circuits of culinary practice.

Senses of belonging amongst the British Nepali diaspora

In Britain, Nepalis are often called Gurkhas, an elite military brigade established for the British Empire in 1816 and still in service today. In the aftermath of World War Two, Gurkha units were absorbed into police and army contingencies where the former British Empire had key posts: India, Burma, Singapore, Hong Kong, and Brunei. As soldiers' wives and children moved to army bases in these locations, a military-based diaspora took shape, creating multiple migration experiences for Gurkha families (Gellner and Hausner 2018). Although the romantic figure of the brave Gurkha has played prominently in military narratives within the British public imagination since imperial times, it was only after 2007 that most Gurkhas were allowed to settle in the United Kingdom (Adhikari 2012). Since then, the British Nepali population has grown to over ten times its previous size in less than ten years, made up primarily of Nepal's indigenous minority ethnic groups ('*janajatis*'). This rise was due to historical recruitment strategies, which targeted specific ethnic groups for the Gurkha armies in a system of colonial racial classifications that designated them as belonging to 'martial races' (Streets 2004). Only four key *janajati* groups out of some 125 ethnic groups and castes of Nepal make up about 70 per cent of Nepalis in Britain: the largest group in Britain, the Gurungs, make up only a little over 2 per cent of Nepalis in Nepal (Adhikari 2012). Therefore, what is considered Nepali food and culture by the British Nepali diaspora itself must be viewed through an ethnic lens of indigenous identity politics that recognises disparities in the demographic representation of Nepali cultures (van Ommen 2018).

However, despite the overwhelming dominance of the few, distinct *janajati* cultures amongst Nepalis in Britain, food practices amongst British Nepalis reflect homogenising aspects of a more singular diasporic consciousness based on Nepal as a country of origin. This process often develops into nationalistic and patriotic discourses. Just as in Nepal, foods considered to form national Nepali cuisine take precedence over particular *janajati* dishes in being consumed across different ethnic groups. Food scholars have presented how the dynamics of embedding symbolic meanings of unity in difference through iconic dishes can be representative of intersections of power and discourses about hybridity and senses of homes as multicultural, cosmopolitan places (Duruz and Khoo 2015; Ku *et al.* 2013). Emerging studies on Nepali diasporas further emphasise fluid senses of belonging when diasporic consciousness becomes contested, rejected, or reconfigured, depending on historical situations (Gellner and Hausner 2018). Nepalis also find themselves belonging to a 'complex diaspora', a term coined by Pnina Werbner (2004) and defined as a dimension of identity based on cultural consumption trends, including food, of geographical regions.

Many British Nepalis dress in the latest Bollywood fashion, sing Urdu songs, and visit Bengali sweet shops for specific occasions, sharing practices common within Britain's complex South Asian diaspora. In the British definition, South Asians are synonymous with the ethno-racial category Asian, different to Asian as a descriptor in the United States, Australia, and other countries where the term usually describes those of East and Southeast Asian heritage. *Janajati* physical features fail to match those widely considered Asian in Britain, and British Nepalis often end up classified as 'Oriental' or mistaken for 'Chinese'. *Janajatis* themselves use problematic descriptors, classifying themselves as 'Mongolian'. Attention to these racial markers adds to scholarship thinking through migration, race, and ethnicity in food studies (Ku *et al.* 2013; Slocum and Saldanha 2016). It reveals how British and Nepali categories of ethnicity and race play parts in facilitating wider senses of belonging that shift between varied senses of being Asian. As Nepalis engage in multiple Asian consumption practices, food and the senses add to these understandings.

Even when not looking 'Asian' in the British sense, finding themselves in a complex diaspora has advantages for Nepalis for integration through markets built by previous South Asian ethnic economies. Before Gurkha army settlement, a large portion of Nepalis migrated to Britain through employment in curry restaurants that served 'Indian' dishes of an adapted cuisine that was already popular in the country. These restaurants marketed themselves through orientalist, romantic stereotypes about Nepal with their iconography and language through imagery of Sherpas, Mount Everest, and Gurkhas. As the Nepali population rose in Britain, their curry restaurants expanded menus by adding popular Himalayan street foods to attract Nepali clients. Diasporic place-making increased as new shops, pubs, bars, and restaurants were established to cater specifically for the tastes and needs of the growing Nepali populace throughout neighbourhoods in London and Southeast England, particularly Aldershot (the town with the highest Nepali population density and headquarters of the British Army). These growing Nepali sites were reasonably proximate: Aldershot is approximately 65 kilometres from central London, while with Nepali neighbourhoods within London, it takes on average an hour to travel from one to another.

This physical connection to cultural food is complemented by photographs and videos uploaded by customers of Nepali food places to YouTube, Facebook, Instagram, and other digital platforms. Nepali food establishments further encourage people to follow their social media pages, use specific hashtags, livestream events, and pinpoint geo-locators. Visuals of these places become constants on British Nepali social media feeds as community members criss-cross between these places for social functions and share their experiences digitally. They create online sensorial circuits of opportunities to vicariously visit and partake of commensal occasions in these real-world places. Senses of intimacy and embodied food practices are enhanced through the co-presence created by virtually sharing food-related activities online. These practices also extend place-making by creating desires to celebrate, affirm, and visit physical locations. Nepali restaurants, shops, bars, and even food trucks are connected through locative functions of social media applications, and their resulting digital cartography draws in places within and outside London through online connectivity. This way, physical distances between Greater London and Aldershot become blurred as spaces are cut and connected virtually, creating a mental constellation of Nepali food places that stretch what maps the city.

Tasting digital food practices through polymedia and Instagrammatics

This research's polymedia approach that investigates digital food practices across several social media platforms is a means to think through pathways in which a digital ecology is expanding to shape both new and existing concepts of diasporic identity and their relationships to place. Polymedia's 're-socialisation of communication media' (Madianou and Miller 2012b: 183) continues the ongoing convergence of earlier theories of mediation and mediatisation (Hepp 2009). It places technological innovation as secondary to social and individual agency in the usage of digital platforms (Herbig *et al.* 2015) and has been applied to describe the integration of digital and traditional media in everyday culinary practices of cooking and shopping (Kirkwood 2018). Although a rich, emerging field of scholarship on digital food-related practices (Rousseau 2012; Lupton 2018; De Solier 2018; Lewis 2018) continues to develop, there are few studies on migration and digital food practices, with some exceptions (Alinejad 2013; Holak 2014; Marino 2017). Researching digital food practices in migrant contexts can further our understanding of the ways in which new sensorial, embodied practices form through transnational communicative methods. For example, domestic spaces extend when Italian family members instruct cooking processes live through Skype and join each other virtually at mealtimes from London to Italy (Marino 2017). Dual senses of home as both Iran and America emerge through online discussions about music videos referencing Iranian eateries and food shops in Los Angeles (Alinejad 2013). This growing field of research on intersections of digital technologies and migrant food worlds is important in uncovering new pathways of both social integration and cultural preservation in a rapidly globalising world. Due to conflict and precarity, as more families and communities become dispersed as refugees or labour migrants, food as a subject of memory, practice, and discovery creates new modes of belonging from connectivity through digital affordances.

To understand multiple senses of home and homemaking better, ethnographic approaches for this research utilises 'Instagrammatics' (Highfield and Leaver 2016) for textual and semiotic analysis. Instagrammatics originally focused on communication through hashtags and other intertextual content for tracking and contained visual languages that included filters, stickers, emojis, GIFs and memes. Since Instagrammatics' original conception, there has been an increase in technologies for adding sound effects to visual material. This research extends Instagrammatics to include aural communication and examines how it is employed by the British Nepali social media ecosystem to follow and influence digital food-related practices. Instagram, YouTube, and Facebook are the most popular social media outlets amongst British Nepalis at the time of research, which may easily be subject to change. They are not necessarily the most popular platforms for all Nepalis, as TikTok is arguably more popular than Instagram in Nepal.

Beginning with YouTube, videos on the topic of Nepali food follow global online food content trends, abounding with restaurant reviews, cooking tutorials, travel documentaries, *mukbangs* (live eating broadcasts), and ASMR (autonomous sensory meridian response) videos. Eating competitions following online viral food challenges and reaction videos by non-Nepalis discovering Nepali food are widely popular. YouTube has become a platform for Nepalis to connect with each other and the rest of the world through a rich range of sensory levels of virtual food experiences. Other public pathways of connectivity and individual sensory gratification are also easily accessible through search functions specific to different social media applications, constructing senses of the depth and breadth of the Nepali global presence through the volume of content. Most content stems from Nepal, its recent diasporas, and Northeast India, as opposed to diasporas of Nepali origin in Thailand and other places. This creates a culturally dominating Himalayan narrative of what foods are quintessentially Nepali. Photographs and videos dedicated to Nepali *momos* ('dumplings') and *dal-bhat* ('lentils with rice', problematically designated as Nepal's national dish when these ingredients are not readily available in many regions of Nepal) fill the internet through searches on Nepali food, signalling that their consumption is part of what makes one Nepali. Nepal's open border with India and its shared ethnic communities on both sides paved the way for *pani puri* ('stuffed pastry shells') and other popular South Asian street snacks to enter the mix of iconic dishes articulating components of a Nepali (and simultaneously a complex South Asian diaspora) food experience.

Extending the map of Nepali food experiences, British Nepalis facilitate an online sensory, visual circulation of diasporic food experience into Europe, a region easily accessible and affordable for many of them. With British passports, British Nepalis do not face the visa restrictions imposed on most other Nepalis in Europe who remain Nepali citizens. The British Nepali YouTube channel 'Momo Sisters' produces reviews of Nepali restaurants around Europe through videos given significant amounts of time to craft and edit. However, despite the channel's public status, encounters of the Nepali food places from its videos are created primarily through private social media networks, especially through ephemeral media. Ephemeral media applications – for example WhatsApp Status, Instagram Stories, and Snapchat – create more immediate senses of connection and awareness of food activities by users uploading content daily. Their predominantly casual forms of visual presentation can inform the analysis of everyday practices through their small, often mundane, moments (Bayer *et al.* 2016), even if short-duration media may also be heavily curated for marketing and creative purposes.

Ephemeral media content tends to be prolific in volume, often uploaded live or within a day, usually in short form with a maximum duration of fifteen seconds. They are no longer as ephemeral or private as scholars previously observed (Jeffrey *et al.* 2019) due to technological developments (they can be archived and re-uploaded, highlighted as permanent albums, or even made longer on platforms like Instagram TV [IGTV]). Nevertheless, most short-duration posts remain spontaneous, uploaded with the pending knowledge that they will disappear within 24 hours, freeing posters from concerns about creating lasting impressions, aesthetics, and data space. Their frequency places them as parts of everyday practice. They become aural and visual sensory channels that fill in the details of what British Nepalis find meaningful. Such information may be easy to miss, depending where focus lies on textual analysis of digital media.

For example, one young woman's Instagram page (@rt_labung), shows photographs of her holiday in Austria. Her Instagram Stories (captured through screenshots by the author before they disappeared) reveal how on the same holiday she chooses to visit a local Nepali restaurant for most of her meals. These culinary encounters would not be gleaned from semiotic readings of her Instagram page, which presents landscapes and architecture in her Austrian holiday photographs. Textual analysis limited to her Facebook and Instagram pages would miss how this young woman actively seeks Nepali and Asian foods

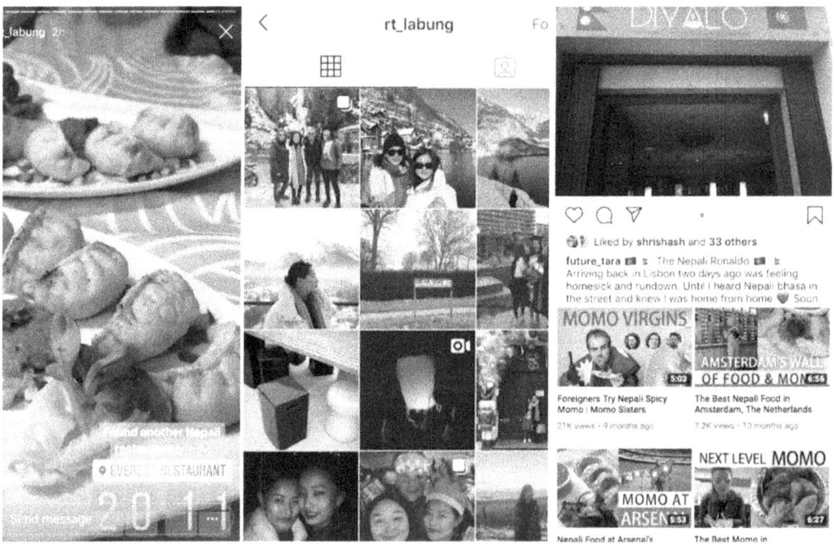

Figure 10.1 Instagram Story feed of Arati Labung (@rt_labung) in Austria contrasts with her fixed Instagram album page, Instagram post of Diyalo Restaurant in Portugal by Tara Manandhar (@future_tara), and screenshots of the Momo Sisters' YouTube channel

whenever travelling. This practice is replicated prolifically by many other British Nepalis, sharing meals on Instagram Stories instead of Instagram pages. Many online platforms convey how common it is for Nepalis to seek Nepali food on their travels. They affirm and celebrate the global Nepali presence and present being Nepali through the repeated love of particular dishes. However, despite signalling common messages, the ways they facilitate the senses differ according to their digital affordances. For example, the ephemeral media of Instagram Stories bring heightened senses of virtually participating in commensal acts due to the casual, recent (and almost or sometimes live), and immersive nature of their uploads, which also allow for immediate comments and responses. Madianou and Miller (2012a, 2012b) show how a polymedia environment even impacts mundane activities with senses of co-presence created through Skype when website cameras give opportunities for families to transnationally instruct and monitor their children as ways of virtual babysitting. Although digital pathways create a range of senses from enhancing embodied experiences to extending everyday practices, their variety does not create hierarchies in social meanings based on upload frequency or other properties of their format. In the variety of digital contents British Nepalis create about Nepali food, sensorial differences become complementary components of discourses about how affection for Nepali food may chart diasporic place-making, invoke memory and nostalgia, and build ties to concepts of home.

In addition to the missing visuals of Nepali restaurants in Austria, videos of Nepali restaurant visits in Lisbon, uploaded in real time through @rt_labung's Instagram Stories, are also missing on her Instagram and Facebook albums. This invisibility contrasts to the easily searchable Momo Sisters' YouTube video of their visit to the same places in Portugal. A slideshow of Lisbon's Nepali restaurants on Tara's Instagram account @future_tara also contrasts with the Momo Sisters' vlog of their Portugal trip with the lack of filters and editing. Her photographs are captioned:

> Arriving back in Lisbon two days ago was feeling homesick and rundown. Until I heard Nepali bhasa *(language)* in the street and knew I was home from home. Soon enough I was eating thali set and ada *(half)* plate [of] momo.

Tara explains she meant 'home' as London, where she was raised with an English mother and Nepali father, without daily Nepali meals. She explains that being homesick on holiday was about a 'shared experience of being dislocated and removed from home' and 'knowing that you crave the same (Nepali) foods'. She recalls that as a child she frequently visited London's Nepali restaurants with her father. Home for Tara refers to British Nepali diasporic food places, instead of Nepal or places with memories of daily Nepali food consumption. Nepali followers on her social media feed post likes as agreement with her caption on missing home but hold other sentiments about what missing home means. For Tara, feeling at home away from home through encounters with food places stretches the sense of the city as a place of opportunity to create diasporic bonds from London to Lisbon.

Adding to senses of home through social media, many Nepali ephemeral media posts capture commensal occasions within households. These reveal practices regarded as private, specific to British Nepalis perception of what is publicly appropriate in Britain in light of historical *janajati* food discourses. For instance, celebrating eating *dhido* ('polenta') and other meals with the hands is commonly seen on British Nepali ephemeral media but remains largely invisible on Facebook and other public digital spaces. Historically, *dhido* has been considered a shameful, peasant staple in Nepali public discourse, and British Nepalis do not eat meals with their hands publicly, even in Nepali restaurants (unlike Nepalis in New York who do).

In contrast, the consumption of beef and pork and spreading *jutto* ('pollution') through sharing food, practices considered taboo for many Nepalis, are highly visible on social media, shifting in the process of migration and growing indigenous people's activism from Nepal. Along with *janajati* relaxation of former rules, food practices acquired through Gurkha family histories of multiple migration are also evident. Small dried fish called *Brunei ko Macha* (fish from Brunei) are incorporated into curried potatoes, pickles, and other Nepali dishes of Gurkha families. This ingredient is unavailable locally in Nepal and uncommon in non-Gurkha kitchens. Fish from Brunei and other popular food items from East and Southeast Asia become parts of the regular cuisines of Gurkha families. Younger generations of these families reflect multiple senses of home through their uploads of dining on cuisines from Hong Kong, Brunei, and Singapore with nostalgic comments about missing homes as those places outside Nepal. Instagram Stories show foods being sent from 'home' in posts of unboxing parcels of snacks sent from relatives in Hong Kong or remembering 'home' by reminiscing about childhood sweets from Brunei discovered in London's Chinatown.

This extension of South Asian identity into other Asian identities is also noticeable in the ways Nepalis seek and incorporate desired foods, influenced by their consumption of Korean pop music, Japanese anime, and other globally popular East Asian media, with further connections made through notions of racial affinity. From impressions of *janajatis* looking Korean to cooking Korean, images of consuming or cooking *kimchi* and other iconic Korean foods are met with admiration. The same admiration extends to posts of consuming Japanese *ramen*. Notions of ethnic belonging, being Nepali and South Asian, and broader senses of being Asian, all float on scales of relevance, depending on situations. Britishness itself, as a component of the UK's Nepali diasporic identity, is performed through social media displaying relationships to particular iconic British foods as forms of integration. This Britishness includes enjoying pints of beer, making English breakfasts, ordering fish and chips, enjoying Sunday roasts, and having British-style Christmas dinners during winter. These dishes may be selected from Nepali senses of what they regard as representative English and British culture, despite many cuisines incorporated into contemporary multicultural British food practices. These images play fantasies of British culture, creating sensory desires of consumption through familiarity with stereotypes. Music also conveys cultural trends and influences specific to Britain. Instagram Stories on Nepali food use AfroBashment, Grime, and other black urban music produced in London. Food images are layered playfully with music, often laced with irony and humour, uncovering multiple messages for those able to read the references.

As food content uploads uncover cultural influences from Gurkha military routes, they also reveal how the roles of Nepali food shift with migration conditions. *Momos*, a street food in Nepal, has played several roles. In Britain, its former lack of availability and time-consuming preparation process led to social gatherings for the making of them. *Momos* became socialised in Britain in ways less common than in Nepal, becoming part of special occasions rather than the everyday. From the streets of Nepal to homes in Britain, *momos* have returned to the streets at affordable rates in the pubs, bars, and restaurants of Nepali neighbourhoods due to growing demand. They also circulate back to homes through innovations in food production, regarded again as convenience food through businesses selling frozen *momos*. The evolution of *momo's* role and diasporic place-making directed by British Nepali food demands can be tracked through the historical progression of their images posted on social media. *Momos* were once part of special occasions in family photograph albums. Gradually *momo* images began being posted frequently, to the point of not being saved on ephemeral media. A large portion of the digital contents around *momos* also includes jokes through numerous memes. Humour is important to take into account the role of foods in communicating senses that are absent physically with senses of smell and taste: one does not laugh at a plate of *momos*, but one can laugh at the ways *momos* are arranged Instagrammatically online. Digital channels can create opportunities for new senses to be created from food practices.

Figure 10.2 Instagrammatics in a collage of Instagram Stories screenshots about Nepali food consumption in the UK, including British meals and ramen

Source: From Instagram accounts @prabinghale by Prabin Ghale, @rozzani by Rosani Thapa Khapangi, @dhawa_a by Dhawa Gurung, @_ramone69 by Ram Gurung, and @punxdidi by Premila van Ommen

A sense of humour is also a tool that serves as a commentary on spatial practices dealing with food, communicated playfully through Instagrammatics. Younger generations draw attention to how their elderly walk in groups embarking on seasonal foraging for *sishnu* ('stinging nettles') and chestnuts in public parks without awareness of British public social etiquette and norms. Elders foraging brazenly become subject to jokes and memes, using semiotic allusions that can only be read with insider diasporic knowledge. Ikea bags, for example, allude to foraging since they are popular amongst British Nepalis for storing *sishnu* to cook as spinach. Stinging nettles are not sold at Nepali food outlets. Footage of elders preparing them leads to YouTube video logs and Instagram Stories sharing this newly acquired knowledge of indigenous food practices through intergenerational transmission. These digital contents are presented through degrees of intimacy and emotional sensibilities, detailing senses of food, the city, and place-making dependent on differing relationships to space and food practices between generations in a diaspora.

Despite gaps, limitations, and differences in usage of digital technologies amongst different age groups, virtual space produces channels to connect generations by their visibility in the community-specific food practices shared on social media. These links are produced by methods of generational inclusion of family and community members for online content creation. Virtual space also creates and extends discourses of local and transnational self-perception from images of food places that reflect diasporic place-making in the city. Made

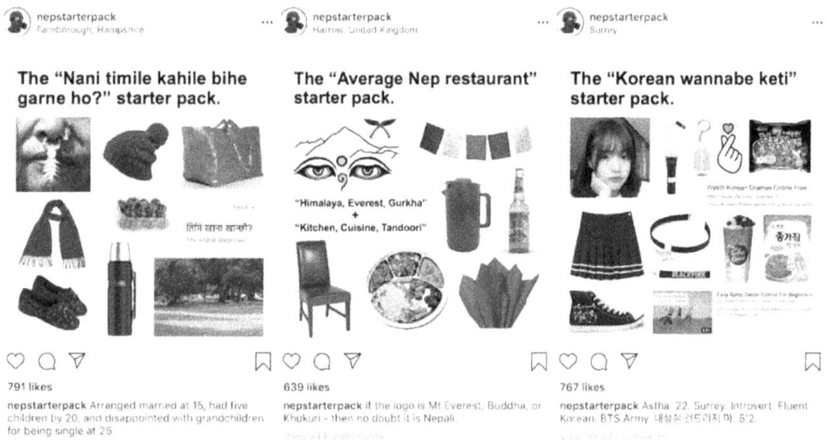

Figure 10.3 Humorous memes playing on internal British Nepali diasporic stereotypes about food practices using Ikea bags for picking stinging nettles, eating kimchi, and other Korean foods

Source: Instagram account @nepstarterpack

visible online, uploads from homes, streets, and businesses reinforce senses of connectivity through enhancing experiences vicariously, allowing viewers to partake in or celebrate dishes, meals, and food-related activities. They also embed online activities to place and generate further place-making by inviting viewers to visit food places. Memories of tastes from multiple senses of home intersect migration histories and nostalgia with contemporary British Nepali food experiences through digital practices. These extend the city to other geographical locations through online networks, bringing global locations of Nepalis and the Himalayas into London in a myriad of sensory experiences enabled by digital food practices.

Conclusion

This chapter has presented glimpses of British Nepali diasporic food-related digital practices and how their contents enhance senses generated through communication circuits, extend socialites, and trace forms of place-making. Their digital contents feed discourses about ethnic, national, and global selves that connect and reconnect with cultures encountered from multiple Gurkha routes of migration. They also display how British Nepali food practices identify with unconnected cultures based on desires produced by East Asian popular culture and notions of racial affinity. This unfolding narrative of selective diversity in food encounters is projected online alongside hegemonic notions of global Nepali cuisine as Himalayan. Food uploads from London's Chinatown connect to Hong Kong as one of the homes out of multiple senses of home for British Nepalis, adding meanings to place in the city. Meanwhile, digital food contents also position new senses of belonging in the ways Nepali food places evolve and shape the landscape of neighbourhoods in Britain. All aspects discussed in this chapter merit more investigation and analysis. They raise questions such as what do digital feeds of home-cooked meals show about changing Gurkha family dynamics, given their histories of men being absent for long periods due to military service? When daughters upload emoji hearts and smiles decorating Instagram Stories of their army fathers cooking meals, what do those stories convey emotionally, and what do they say about migrant conditions shifting gender roles? How do digital feeds produce discourses of new configurations for identity formation, and how do they affect sociality by embedding new relationships between food, the body, and place? Digital methodologies investigating such questions need to move away from focus on singular social media platforms to polymedia environments of multiple integrated digital social media systems. They need to take into account technological developments in social media application convergences, communicative languages, aesthetic trends, and temporalities. Grounding these approaches that explore connections between embodied and digital practices to place, we can pay deeper attention to ephemeral media and discover social meanings through fleeting moments of taste.

References

Adhikari K. P. (2012) *Nepalis in the United Kingdom: An Overview*. Reading: Centre for Nepal Studies UK.

Alinejad D. (2013) 'Locating Home in a "Digital Age": An Ethnographic Case Study of Second-Generation Iranian Americans in LA and Their Use of Internet Media', *Iranian Studies* 46 (1): 95–113.

Bayer J. B., Ellison N. B., Schoenebeck S. Y. and Falk E. B. (2016) 'Sharing the Small Moments: Ephemeral Social Interaction on Snapchat, Information', *Communication & Society* 19 (7): 956–977.

De Solier I. (2018) 'Tasting the Digital New Food Media', in Lebesco K. and Naccarato P. (eds) *The Bloomsbury Handbook of Food and Popular Culture*. London: Bloomsbury: 54–64.

Duruz J. and Khoo G. C. (2015) *Eating Together: Food, Space, and Identity in Malaysia and Singapore*. New York: Rowman & Littlefield.

Gellner D. N. and Hausner S. L. (2018) *Global Nepalis: Religion, Culture, and Community in a New and Old Diaspora*. New Delhi, India: Oxford University Press.

Hepp A. (2009) 'Differentiation: Mediatization and Cultural Change', in Lundby K. (ed) *Mediatization: Concepts, Changes, Consequences*. New York: Peter Lang: 135–154.

Herbig A., Herrmann A. F. and Tyman A. W. (2015) *Beyond New Media: Discourse and Critique in a Polymediated Age*. Lanham, MD: Lexington Books.

Highfield T. and Leaver T. (2016) 'Instagrammatics and Digital Methods: Studying Visual Social Media, from Selfies and GIFs to Memes and Emoji', *Communication Research and Practice* 2 (1): 47–62.

Holak S. L. (2014) 'From Brighton Beach to Blogs: Exploring Food-Related Nostalgia in the Russian Diaspora', *Consumption Markets & Culture* 17 (2): 185–207.

Jeffrey H. L., Ashraf H. and Paris C. M. (2019) 'Hanging Out on Snapchat: Disrupting Passive Covert Netnography in Tourism Research', *Tourism Geographies*, 1–18.

Kirkwood K. (2018) 'Integrating Digital Media into Everyday Culinary Practices', *Communication Research and Practice* 4 (3): 277–290.

Ku R., Manalansan M. and Mannur A. (2013) *Eating Asian America: A Food Studies Reader*. New York: NYU Press.

Lewis T. (2018) 'Digital Food: From Paddock to Platform', *Communication Research and Practice* 4 (3): 212–228.

Lupton D. (2018) 'Cooking, Eating, Uploading: Digital Food Cultures', in LeBesco K. and Naccarato P. (eds) *The Bloombsury Handbook of Food and Popular Culture*. London: Bloomsbury: 66–79.

Madianou M. and Miller D. (2012a) *Migration and New Media: Transnational Families and Polymedia*. London: Routledge.

Madianou M. and Miller D. (2012b) 'Polymedia: Towards a New Theory of Digital Media in Interpersonal Communication', *International Journal of Cultural Studies* 16 (2): 169–187.

Marino S. (2017) 'Digital Food and Foodways: How Online Food Practices and Narratives Shape the Italian Diaspora in London', *Journal of Material Culture* 23 (1): 263–279.

Rousseau S. (2012) *Food and Social Media: You Are What You Tweet*. Lanham, MD: Altimira.

Slocum R. and Saldanha A. (2016) *Geographies of Race and Food: Fields, Bodies, Markets*. London: Taylor & Francis.

Streets H. E. (2004) *Martial Races the Military, Race, and Masculinity in British Imperial Culture, 1857–1914*. Manchester: Manchester University Press.

van Ommen P. (2018) 'New Gurkha Kitchen: Eating Nepali in England', *La.Lit: A Literary Magazine* 9: 22–31.

Werbner P. (2004) 'Theorising Complex Diasporas: Purity and Hybridity in the South Asian Public Sphere in Britain', *Journal of Ethnic and Migration Studies* 30: 895–911.

Part III

Disrupting and re-imagining

11 A taste for *tapatío* things

Changing city, changing palate

Melissa S. Biggs

The street I live on in Guadalajara, Mexico, is noisy. My street also smells. I can fairly accurately tell you the time of day based on the sounds and smells on my street. Clanging of a cowbell and faintly sour or rancid odour? 8:30am, garbage pick-up. A tin whistle? The knife sharpener, his whetstone mounted on the bicycle he pedals by slowly. Jingling bells? 6:00pm, the ice cream vendor. The rich scent of frying pork fat mixed with the faint trace of wood smoke? Midnight, when one of my neighbours prepares *chicharrones* (pork rinds); sometimes bits of ash from the fire blow into my outdoor wash area. At 4:30 or 5:00am, the damp, vegetal aroma of steamed corn dough and the clanging pot lids and shouts of the family across the street loading their *tamales* to sell in the *centro*. More recently, the teasing banter of teenagers gathered at the snow cone and snack stand my neighbour runs out of her living room tells me that it's around 8:30 or 9:00 at night. These scents and sounds mark the changing hours of the day. Cities, however, are dynamic. I returned from a brief absence to find the family and the aroma of their tamales gone, the house renovated. This chapter considers changes in the sensorial experiences of eating in Guadalajara as markers of broader economic and social shifts in the city.

Background and methods

I arrived in Guadalajara in August 2016 to begin an ethnographic project exploring culinary tourism in the state of Jalisco. In late 2017, I was invited to participate in the citizen's advisory board for the food supplement of one of the city's daily newspapers. The idea for this chapter began forming during conversations that took place during the editorial board meetings I attended biweekly from February through December 2018. The composition of the board changes yearly; it is meant to represent a broad sample of the reading public. The group in which I served included chefs, café owners, and others employed in the food industry, as well as business people and a self-described housewife. Our ages ranged from late 20s to early 60s. The majority were *tapatíos*, as natives of Guadalajara are known; others came from Mexico City, Sinaloa, and Baja California Norte. Frequently, discussion at the board meetings turned to changes in local foodscapes: aromas, tastes, and textures remembered and no longer available and the introduction of new culinary offerings and flavours, some more welcome than others.

It was at a board meeting that I first heard someone refer to a 'palate' particular to the city. A member whose husband is from Mexico City mentioned the balance she struck in the meals she plans for her family between her *paladar tapatío* – 'tapatio palate' – and her husband's *chilango* (slang for people from Mexico City) tastes. She prepares some recipes shared by her mother-in-law, satisfying her husband's cravings and introducing them to their children. In this way, the children learn to appreciate both tapatio and chilango foods.

For this work in progress, I combine my time as a participant observer on the advisory board with observations made over the last three years and formal and informal conversations with culinary professionals, my neighbours, and other Guadalajara residents. The basic questions I ask are: What does change taste like? What are its smells, its sounds? I am also interested in thinking about methodologies, and what sorts of writing best convey the complicated and layered 'inquiring relationships' (Heldke 1992) upon which investigations into food, senses, and place rely, including the sensory experiences themselves.

David Sutton coined the term 'gustemology' to describe 'approaches that organize their understanding of a wide spectrum of cultural issues around taste and other sensory aspects of food' (2010: 215). Recent work by Susanne Hojlund (2015) builds on Pierre Bourdieu's (1984) notion of habitus to develop a theory of taste as a social sense. She points out that Eurocentric ideas of taste typically consider it an individual matter, attending to 'the privacy of the mouth, and the subjective'. By shifting attention from eating to tasting, we come to understand tasting itself as a practice. This, she argues, enables us to 'understand how ideas of food quality and preferences for certain foodstuffs are brought into the social and thereby being object for others and possible to share'. To implement the sort of analysis Hojlund calls for, I take as a starting point C. Nadia Seremetakis's expanded definition of commensality as 'the exchange of sensory memories and emotions, and of substances and objects incarnating remembrance and feeling' (1996: 37). As ethnographers, we cannot 'directly access the "collective" memories, experiences or imaginations' of other people; however, she suggests that by 'attuning our bodies, rhythms, tastes, ways of seeing and more to theirs, begin to make places that are similar to theirs, and thus feel that we are similarly emplaced' (Pink 2008: 193). Seremetakis's 'reflexive commensality' implies the sort of emplacement Sarah Pink advocates.

The Spanish word *convivir* expresses that connected sense of being together. While the official definition of *convivir* according to the Academía Real de la Lengua is 'to live in the company of another or others', it conveys the kinds of exchanges Seremetakis presents and the attuned awareness of presence described by Pink.[1] The act of *conviviendo*, *convivencia*, provides a means to explore and begin to define a place-specific palate. An important aspect of convivencia is the *sobremesa*, the talk around the table after a shared meal. The editorial board meetings opened with a breakfast cooked onsite and sometimes included foods prepared by board members using recipes appearing in that week's issue of the newspaper or their family's version of a traditional dish. Conversation about the shared food – what sort of chilli was in the salsa served with our eggs, adjustments made to the printed recipe, how preparations of the dish differed among board member – often prompted the direction of the discussion that followed. Convivencia also occurs in shared spaces such as market stalls and corner taco stands and during community celebrations for patron saints, places where 'discursive meaning-making activities of daily life are not separate from sensuous, material life' (Hayes-Conroy and Hayes-Conroy 2010: 1280). I argue that convivencia provides a framework to understand the ways in which tapatios both note and manage change.

Tastes of Guadalajara

In the wake of the 2010 inscription of Mexico's gastronomy into the United Nations Educational, Scientific, and Cultural Organisation (UNESCO) Representative List of the Intangible Cultural Heritage of Humanity, the Mexican government launched a number of initiatives intended to foster and promote gastronomy as a tourist attraction and a motor for sustainable development. The proposal submitted by the Mexican delegation specifically named 'the Michoacán Paradigm' as the case study for UNESCO consideration. The central western state of Michoacán is widely lauded for its culinary traditions; its neighbouring state of Jalisco, while famous as the centre of tequila production, is not generally noted in Mexico for its exceptional traditional food. The signature dishes often associated with the state, such as *tortas ahogadas* ('drowned sandwiches'), made with chopped pork known as *carnitas* stuffed into a dense, faintly sour bread with an especially crunchy crust called a *birote*, then doused with both a plain tomato sauce and a chilli sauce; *birria*, a chilli stew usually made with goat, lamb, or beef; *pozole*, a hominy-based chilli stew typically made with pork; and *carne en su jugo*, beef cooked in a *tomatillo* ('Mexican husk-tomato') sauce and served with brothy beans garnished with bacon, are considered simpler both in preparation and in flavour than the moles and tamales of other regions. I wondered how government efforts to increase food tourism would fare in a state with a relative lack of culinary cachet.

While Jalisco might not be one of the first places recommended to those seeking complex traditional dishes, Guadalajara was already known nationally for its chef-centred restaurants. I arrived as the city's food scene was moving from national to international recognition. Chef Francisco Ruano, a Guadalajara native, opened his restaurant, Alcalde, in 2013. By 2017, the restaurant had made its inaugural appearance on the San Pellegrino list of 'Latin America's 50 Best Restaurants'. In June 2019, it appeared on the list of the 'World's 50 Best Restaurants'. Between 2016 and 2019, articles about the city's food scene appeared in the *New York Times*, *Travel and Leisure*, and *Food and Wine*, among others. Despite this, Jalisco lacks the culinary tourism infrastructure found in other locations, such as Mexico City or Oaxaca. Typical food tourism activities available in other cities, such as cooking classes and food market tours, are scarce. I concentrated on attending to the ways Guadalajara residents experience and talk about food to better understand what residents found unique and delicious about their city's food and what they wanted visitors to know about it.

Though most middle-class Mexicans in urban areas have access to a wide variety of ingredients and cuisines, their food preferences generally remain regionally specific. For everyday eating, people favour the products and preparations typical of their areas. For example, in Guadalajara, the black beans preferred in the centre and south of Mexico rarely appear in home cooking or on the menus in market stands or the small home-style eateries known as *fondas*; they are considered *feos* ('ugly'), not appetising. At one of the early meetings of the newspaper advisory board, a member lamented the tapatio proclivity to soak food in sauce, calling it 'the sin of the tapatio . . . we're the children of *salsa*' ('*Sí, somos bien soperos*' ['We really like to drown our food']), agreed another. I recalled the exasperated complaint of a friend from Mexico City, now living in Guadalajara: 'It's the only place in Mexico where you have to use silverware to eat street food; they drench everything!' Others chimed in with more peculiarities of the Guadalajara palate: a fondness for the acid tastes of lime and vinegar; salt, sweet, and savoury combinations; a love of chilli. Some members equated these preferences with a palate needing 're-education' in order to appreciate less assertive flavours. Others rejected the notion that palates needed reforming, citing the need to 'respect [local] customs'.

I began noting similar comments about tapatio preferences made in everyday conversation and interactions. That led to a short project about perceived differences between the preferred bread of Guadalajara, the dense birote, and that of Mexico City, the *bolillo*, with its crisp crust and light crumbs. While their criteria for evaluating bread differed, those I interviewed in both cities lamented that they could no longer reliably find bread they considered 'good'. Along with the pleasure that food brings, several of the factors that Abby Wilkerson (2016) cites as constituting 'good food' – including 'economic, logistical, nutritional, temporal and political considerations, along with individual desires and preferences' – surfaced in my interviews. Consumers cited 'freshness' and 'flavour' as qualities lacking in most readily available bread. They cited the pressures of time causing them to choose to buy bread at supermarkets or convenience stores rather than bakeries and the reliance of industrial bakeries on machinery and artificial flavours. People noted the disappearance of local bakeries and the street vendors who sold bread in the early morning or late afternoon.

The bakery owner I interviewed in Mexico City stated that younger people had acquired a taste for dough treated with conditioners and flavour enhancers. A baker at Panadería del Río, a Guadalajara bakery where the dough for birotes is still mixed manually, explained that 'other places use machines, here everything is by hand. . . . [I]t gives a different taste. Not everyone likes it'. He mixes the dough in a long wooden trough, rinsed but not scrubbed down at the end of the baking shift. Preparing foods using traditional wooden and clay utensils is believed to imbue flavour, partially through the texture and taste of the materials themselves and partially through the residues left by previous foods preparations. None of the bakers wears gloves when handling the dough. Bakers and consumers alike commented euphemistically on the 'saltiness' this contributed to the dough.[2] Hand-mixing gives the crumb of the bread a denser texture; hand-shaping means that the rolls are not perfectly uniform in size and shape. Panadería del Río bakes its breads in brick ovens, also contributing to a less uniform product: some of the rolls come out more *doraditos* ('golden', referring to crust colour and crunch) than others. Shoppers comment to each other and to vendors about which they prefer.

Changing city, changing tastes

Following the bread project, I broadened my inquiries into tapatio tastes, thinking through the ways in which residents connect – or perhaps do not – their experiences of changes in the city with changes in food availabilities and preferences. Applying the concept of a 'palate' peculiar to a place intrigues me, especially when considering an urban setting as spread out and diverse as Guadalajara, the second largest city in Mexico. Initial population growth began as economic policy at the national level shifted from the protectionist policies of the 1930s to 1950s to the import substitution on the 1950s to 1970s, and people moved from rural to urban areas seeking jobs. In the late 1960s and early 1970s, the city supported the creation of the Guadalajara Industrial Zone – similar to the Border Industrialisation Zone that brought maquiladoras to the Mexico-USA border – attracting companies such as IBM to the city. Hewlett Packard arrived in the early 1980s, followed by other multinationals. But in the 1990s, competition from China saw many of the companies that had come to Jalisco leave. Additionally, other states in Mexico became more competitive.

The early 2000s brought massive layoffs, resulting in a transformation of the existing industrial sector to one oriented towards software, informatics, and telecommunications (Partida Rocha 2010). At roughly the same time, the city sprawled outward, increasing in size by a little more than 380 per cent between 1980 and 2010 (Padilla *et al.* 2016). In addition to the city itself, the Guadalajara Metropolitan Area (GMA) now encompasses seven other municipalities. The total population of the GMA reached five million in 2017 (IIEG 2017a), 62 per cent of the total population of the state (IIEG 2017b). As in many Latin American cities at the time, the population of the urban centre shrank. Wealthier families moved into housing developments called *cotos*, walled and gated private communities in the city's outskirts. Now, nearing the end of the second decade of the 2000s, Guadalajara aims to become Mexico's first 'smart' city. Efforts are underway to 'rehabilitate' and 'reclaim' the deteriorated city centre by converting buildings into condominiums and co-working spaces. Luxury apartments are under construction in several of the core central neighbourhoods, including Santa Teresita, where I live. Heralded for its traditional market, *taquerías*, and restaurants, it now houses a pho restaurant, multiple sushi outlets, pizza franchises, and stores stocking kombucha and quinoa.

Before moving to Santa Teresita, I stayed for several weeks with friends who live in one of the gated cotos. Neither public transport nor ambulatory vendors can enter. The sights, smells, and sounds typical of my street in Santa Teresita do not pass into the coto. Regulations in the coto prevent people from regularly frying pork rinds for sale or setting up temporary food stalls based in their homes when they need quick cash. There is no nearby neighbourhood food market, nor is there a temporary outdoor market, called a *tianguis* in Mexico, within easy walking distance. All shopping occurs outside the coto. Small commercial centres including a mix of convenience stores, fast food chains, and small businesses selling prepared foods such as packages of grilled meats, tortillas, and salsas primarily for take-out cluster around the entrances to the gated communities. Large one-stop supermarkets like the nearby Walmart supply most household needs.

Walmart first arrived in Mexico in the early 1990s, in partnership with a national chain that it later purchased. The opening of multinational chains like Walmart, with international supply chains, marked one change commented on both in the advisory board meetings and in interviews. A woman in her mid-30s told me that she remembered when the first multinational supermarkets arrived, after the merger: 'The apples were so big and red. It gave people a different idea of what an apple should look like and taste like'. Consumers began to prefer the imports to the smaller, tarter apples grown in Chihuahua and other parts of Mexico. People connected these changes to 'lost' tastes for things like *quelites* – wild greens, often harvested from small corn and vegetable plots and consumed or sold – and local fruits, like *capulines*, sometimes called 'wild black cherries'; *guamúchil*, a pod that contains shiny black seeds surrounded by an edible pulp; and *cocuixtle*, a bromeliad that produces a fibrous, sour fruit. 'People have forgotten about those foods, or they think they're for rural people', one of the advisory board members said. These preferences filtered down into the more local markets as well. Only one vendor in the weekly food tianguis at which I shop reliably sells quelites. 'No one asks for them', she told me. Changes in city infrastructure eliminated sources of foraged foods. One of my neighbours, a woman who lives in the house that once belonged to her grandparents, reminisced about her childhood, when she was sent to the nearby river banks to gather plants that they needed for food or home remedies: 'We would swim in the river before we went home'. Like most of the rivers and streams that flowed through Guadalajara, the river to which she referred has long since been channelled and paved over as part of urban expansion projects.

People also commented on changes in the way food is presented in both the traditional indoor markets and the tianguis, remarking particularly on the restructuring of the Mercado Corona, a large market located in the city centre. A fire in 2014 destroyed 90 per cent of the structure. When the rebuilt market reopened in 2016, 'they really cleaned up', an interviewee told me. 'Everything was modernised'. Previously, produce sold in the market 'had the dirt it came out of the ground with', but now it is scrubbed and sometimes even packaged like supermarket produce. Mobile street vendors still sell prepared foods from bicycles or push carts, but they are being replaced by vans or pickup trucks that move much more quickly and cover more territory. The faster pace means fewer exchanges among shoppers and between vendor and buyer.

The value of food

The lack of connection between food production and consumption concerned the editor of the features section that includes *Buena Mesa*. She viewed addressing this lack of connection to raw ingredients and local products as part of the section's mission. Stories during the time that I served on the board often highlighted a food trend by focusing on local versions. For example, a story about fermented foods featured the traditional fermented corn beverage *tejuino*, as well as Mexican chefs producing sauerkraut and other fermentations not typically found on Mexican tables. The role of the newspaper is 'to bring the topics to the table' and allow readers to make up their own minds. People want to know how to choose quality ingredients and prepare healthy meals but often do not know where to begin: 'People don't know that they don't know', she stated. They care about environmental and social issues implicated in food choice. She cited an issue about nixtamalisation and landrace corn published as a result of advisory board discussions and suggestions. Nixtamalisation refers to the process of soaking and cooking dried corn in an alkaline solution in order to remove the hulls from the kernels, allowing it to be ground and increasing its nutritional value. The issue described how to distinguish tortillas made from nixtamalised landrace corn from tortillas made from industrial corn flour through appearance, texture, and flavour. Articles introduced local producers, discussed why products made from landrace corn often cost more than industrialised ones, and provided information about nutritional and environmental benefits offered by landrace corn that consumers could balance against the additional cost.

To speak of a 'tapatio palate' implies tastes that are shared. The public who reads the food section of the newspaper, watches specialty food programming, or uploads photos of meals to social media represent particular sectors of the consuming public, those with the time, interest, and resources to consider food choices and to eat as a form of recreation, not just sustenance. Those with fewer resources think about food differently. A community nutritionist who works in poor and working class sections of the metropolitan area explained the calculus the residents she serves consider when feeding their families. While they might prefer the taste of fresh produce, in homes that lack refrigerators, shelf-stable goods mean less chance that food will spoil before it is eaten. People with limited incomes are less willing to take a chance on unfamiliar foods: if their family does not like it, that is money wasted. Also, emphasising nutritional value sometimes confuses clients. She described one mother who chose fortified and sweetened juice-based drinks rather than whole fruit for her children because the packaging on the juice provided information about its vitamin and mineral content, information not readily available about fresh fruit. She also knew that her children would drink the beverages; she was not sure they would eat fruit.

However, economic considerations do not necessarily trump flavour preferences. While the Guadalajara area is recognised as a manufacturing and technology centre, Jalisco itself is one of the top Mexican producers of produce for export. Cherry tomatoes are a popular export crop and also appear in the tianguis and on restaurant menus as specialty ingredients in salads or pasta dishes. They typically cost more than other tomato varieties. The community nutritionist regularly visits the metropolitan food banks. On one visit, she observed piles of red and yellow cherry tomatoes. The manager told her that the people served by the bank are unaccustomed to eating fresh tomatoes except in salsas or broths and found the cherry tomatoes too sweet for those purposes. The nutritionist, a Guadalajara native, admitted that she also finds it difficult to incorporate uncooked vegetables into her diet: 'We didn't eat that way when I was growing up'. Like her clients, she struggles to change eating habits learned in her family, even though her studies lead her to believe that changing them improves her diet.

The lure of novelty and the pull of tradition

The post–North American Free Trade Agreement (NAFTA) influx of sugar and industrialised corn- and soy-based snacks originating from the United States introduced new tastes and created what a nutritionist I interviewed described as an 'altered expectation of flavour'. Processed snack foods cause 'a change in palate perception', she told me. 'People start to expect things that are saltier, or sweeter'. Benjamín, twenty, reminisced about *turrón rosa*, a traditional candy with a texture similar to taffy that was commonly sold outside schools and at public events like patron saint festivities, served with a squeeze of lime and a sprinkle of powdered chilli. Vendors displayed it in blocks, and, as Benjamín described, 'they would cut the amount you wanted with machetes, THWACK! You could ask for one peso, or three pesos. Sometimes if I wanted to show off, I would ask for ten pesos, to share with my friends'. 'That's right', Isaak, 39 years old, added. 'They would always bring the machete down hard when they cut it'. Neither could recall the last time they had seen a turrón vendor: 'It's not what people want anymore'. Benjamín also noted that he no longer saw vendors selling freshly fried potato chips, a common sight when he was younger. He wondered whether the plain cellophane bags were not as appealing as the shiny, bright packaging of mass-produced snacks. Both he and Isaak cited advertising by large companies as a factor in the decline of traditional snacks. They believed that the packaging and association with cartoon characters or other figures from popular culture heightened their appeal to children, particularly.

For many people, they observed, image matters more than flavour. They attributed this to the effects of social media. 'People want to be somewhere "nice" so they can post it to Facebook', said Isaak, using the English 'nice'. 'Yeah, it's all about the concept', Benjamín agreed. This carries over into the transformation of space. Upscale, more photogenic cafés and bars are pushing out older establishments. Isaak, a musician, cited the decrease in the number of *centros botaneros* ('bars'), called so because of the practice of serving a *botana* ('snack') when patrons order a drink. Typical snacks are pickled chillies and carrots, shrimp broth, and tacos with various fillings. He had recently performed at one of the few left in the city centre. He stated:

> They're not in style anymore; people think they're for old people, or for getting drunk. It's true there isn't much atmosphere, maybe some plastic tables with soda or beer logos on them. But the food is good, and the beer is cheap.

Benjamín noted the disappearance of small family-owned restaurants and cafés, increasingly replaced by sleeker places offering cappuccinos and bagels.

Not all allegiance to more local forms of eating is lost. Street *taquerías* emerged as one site interviewees believed would withstand change. Isaak contrasted taquerías to the cantinas: 'Everyone has a favourite place to go, it doesn't matter where you live or who you are. You can see the fanciest car pulling up to the smallest, grimiest taquería'. Good food, in all its aspects, outweighs social media image-making. When talking about what makes a taco delicious, people describe not only the tastes, but the aromas, textures, and sounds and the bodily experience of eating at a taquería: the scent of sizzling fat rising off the griddle making their mouths water; the visual appeal of salsas made from diced and ground vegetables, herbs, and chillies; the *rat-a-tat-tat* of the *taquero*'s knife on the wooden block as he chops meat; the crunch of perfectly fried tripe; the pleasure of crowding together around the kiosk; the skill needed to eat with one hand while balancing a plate and holding a bottled beverage in the other. Within 'a meta-narrative of loss' (Srinivas 2006: 193) that marks the rapid transformation of urban spaces and the loss or replacement of familiar tastes, taquerías provide a site for convivencia that both discursively and materially resists homogenisation and sterility.

In this chapter, I aimed to begin the construction of a 'cultural chronology' (Paterson *et al.* 2016) – the array of tastes, aromas, and sounds available and preferred across time that contribute to a particular set of preferences – in this case recognised as a 'tapatio palate' – and thus to an understanding of one of the ways in which urban change registers for Guadalajara's residents. Convivencia served both as an ethnographic method employed to learn about these preferences and as a construct through which to analyse tapatio efforts to attend to and manage change.

Notes

1 The title of a special issue of the *Anthropology of Food* devoted to ritual foods in Mexico conveys this use of the word: '*Dar de comer para convivir*', which I loosely translate as 'sharing food in order to share time together'. It describes, for example, the act of placing food on the offerings for Days of the Dead, inviting the departed to join their loved ones.

2 My colleague Anne Johnson told me that during her fieldwork in the state of Guerrero, people told her the saltiness of the local tortillas was the result of the women holding their hands under their arms periodically before patting out the tortillas.

References

Bourdieu P. (1984) *Distinction: A Social Critique of the Judgement of Taste*. Cambridge, MA: Harvard University Press.

Hayes-Conroy J. and Hayes-Conroy A. (2010) 'Visceral Geographies: Mattering, Relating, and Defying', *Geography Compass* 4 (9): 1273–1283.

Heldke L. M. (1992) 'Recipes for Theory Making', in Curtins D. W. and Heldke L. W. (eds) *Cooking, Eating, Thinking: Transformative Philosophies of Food*. Bloomington: Indiana University Press: 251–265.

Hojlund S. (2015) 'Taste as a Social Sense: Rethinking Taste as a Cultural Activity', *Flavour*, https://flavourjournal.biomedcentral.com/articles/10.1186/2044-7248-4-6.

Instituto de Información Estadística y Geográfica (2017a) 'Nace el habitante 5 millones en el AMG', *Strategos: Revista Digital del Instituto de Información Estadística y Geográfica*, https://iieg.gob.mx/strategos/alcanza-area-metropolitana-de-guadalajara-los-5-millones-de-habitantes/.

Instituto de Información Estadística y Geográfica (2017b) 'Poblacion en Jalisco', *Strategos: Revista Digital del Instituto de Información Estadística y Geográfica*, https://iieg.gob.mx/strategos/portfolio/poblacion-en-jalisco-2017/.

Padilla M., Rizo A., Morales O., Contreras L. M., Córdova W., Peregrina E., Lanhart J., Hernández V., Arriaga C., Saez M. and Ocegueda M. (2016) *Analisis del Mercado Inmobilario de la Zona Metrapolitana de Guadalajara*. Guadalajara, Jalisco, Mexico: Cuadra; The United States Agency for International Development; And the Institute for Transportation and Development Policy.

Partida Rocha R. (2010) 'De-industrialization of the GMA: The Case of the Electronics Industry in the Face of Chinese Competition', in Audirac I. and Arroyo Alejandre J. (eds) *Shrinking Cities South/North*. Mexico City: Universidad de Guadalajara Centro Universitario de Ciencias Económico Administrativas; PROFMEX; And Juan Pablos Editor, SA: 217–234.

Paterson M., Dodge M. and Mackean S. (2016) 'Introduction: Placing Touch Within Social Theory and Empirical Study', in Paterson M. and Dodge M. (eds) *Touching Space, Placing Touch*. London: Routledge: 1–28.

Pink S. (2008) 'An Urban Tour: The Sensory Sociality of Ethnographic Place-Making', *Ethnography* 9 (2): 175–196.

Seremetakis C. N. (1996) 'The Memory of the Senses, Part II: Still Acts', in Seremetakis C. N. (ed) *The Senses Still: Perception and Memory as Material Culture in Modernity*. Chicago: University of Chicago Press: 23–41.

Srinivas T. (2006) '"As Mother Made It". The Cosmopolitan Indian Family, "Authentic" Food and the Construction of Cultural Utopia', *International Journal of Sociology of the Family* 32 (2): 196–221.

Sutton D. E. (2010) 'Food and the Senses', *Annual Review of Anthropology* 39: 209–223.

Wilkerson A. (2016) 'Judging, Tasting, Knowing "Good" Food', *Food, Culture & Society* 19 (2): 223–226.

12 The foodie *flâneur* and the periphery of taste in Bucharest's street food scene

Monica Stroe

Situated midway between three popular neighbourhoods, Obor Market is the largest and one of the oldest marketplaces in Bucharest, Romania. Drawing its name from its original function as a cattle market, Obor now focuses on fresh affordable produce and food products but also includes a virtually unrestricted range of merchandise, such as cheap home improvement items, kitchenware, and clothes. Notwithstanding its several reorganisations over the past century, it still summons the image of a bazaar with many improvised stalls and informal appendices (street curb sellers). It is known as a site for affordable purchases and is typically frequented by lower-income shoppers from more peripheral neighbourhoods.

Prized for its quaint, 'authentic' atmosphere and its open-air eateries in particular, the market has become so attractive to middle-class patrons in recent years that it became a trending hashtag on Facebook and Instagram (#obor, #myobor). The weekend market trips, with their ritualised stops at the market's bodegas – where people crowd to eat a meat dish called *mici* – have started to be increasingly documented on social media. Obor's popularity peaked when it became a destination for urban safaris – afforded by tourists and middle-class locals, framed as 'alternative' or 'bohemian' tours of Bucharest. Two female bloggers initiated in 2018 perhaps the most notorious of these tours under the mock-French language title *Tour d'Obeur*. Charging €25 for market 'insider' access and 'know-how', it was cancelled and deleted from Facebook after it came under attack from other Obor enthusiasts who accused the organisers of commodification and gentrification. Some groups based their accusations on an implied symbolic ownership of Obor: one social media comment read 'Get your hands off Obor'. This apparent conflict over monopoly rights to the market experience led me to become interested in the consumption of *mici*, the quintessential street food that makes such places magnetic to a segment of the middle class interested in the acquisition of eclectic food tastes and experiences: the foodie *flâneurs*. Their multi-sensory exploration of the urban foodscape is consistent with the nineteenth-century urban character of the *flâneur* described by Walter Benjamin (1996), whose local and contemporary expression I attempt to describe in the course of this chapter.

I focus on Obor Market as the key site of consumption and on the celebration of Labour Day at the market in local foodie circles in 2018 and 2019. I read this feast as a public arena for the articulation of taste. I analyse the foodies' patronage of Obor as an act of appropriation, for which they mobilise privileged cultural capital to extend the material boundaries of class to working-class foods and consumption sites. I conducted ethnographic research of the local middle-class foodie scene between 2017 and 2019 with a focus on the emerging street food phenomenon. I supplement the ethnographic data and interviews with a digital ethnography that examines foodie discourses and representations of the street food scene (hashtags, online events, and audiovisual material).

Food and class during postsocialism

Mici represent iconic working-class food. Currently priced at about €0.60 apiece, they are affordable, freshly grilled minced meat rolls (generally a mix of fat pork and beef) with a dominant garlic taste that are served with mustard. Draft beer is its typical pairing. Historically defined as street food in Romania, *mici* are available particularly around open-air markets or bus terminals and typically eaten off cardboard plates, using toothpicks as cutlery. Romanians commonly purchase *mici* from a hole-in-the-wall type of kiosk and – in most cases – consume them standing up. Raw *mici* patties are available for home preparation in supermarkets, but consumers comparatively rarely prepare *mici* themselves as they are typically reserved for picnics, barbecues, and other outdoor events. Throughout Romania, *mici* are historically central to the festive socialist consumption rituals on Labour Day (1 May). During state socialism, Labour Day was a bank holiday, celebrated with state authorities–coordinated street parades, which were obligatorily attended by brigades of workers and students. Factory unions would organise so-called popular gatherings and reward workers with barbecues where *mici* were the central feature. Outside the official celebrations, people would have outdoor barbecues that heavily featured *mici* too.

After the end of state socialism, the working class was downgraded in economic and symbolic terms. Privatisation, de-industrialisation, the emerging service economy, the dissolution of unions, and the flexibilisation of labour were among the factors that eroded workers' identities during postsocialism and limited, alongside their symbolic capital, their participation in the emerging consumer society (Berdahl 2005; Kideckel 2010). A mineworker interviewed by David Kideckel in the early 2000s illustrated postsocialist economic hardship through the inability to afford enough meat and lard (2010: 154). Having lost their legitimacy in the public discourse, the tropes associated with the socialist working class, including Labour Day, underwent a process of re-signification, either as negative metaphors or as resources for ridicule in advertising. Labour Day lost its state-propagandistic objectives but did remain a bank holiday celebrated in private, typically with outdoor gatherings, such as picnics or trips to the seaside or the mountains.

Simultaneous to this demise, a postsocialist middle class was emerging, stimulated by the rise of the IT industry, creative industries, and entrepreneurship. The new middle class was particularly eager to define itself in aspirational terms, through lifestyle choices such as leaving the city for detached family homes in suburbia; holidays abroad; and taking up particular sports, such as tennis, skiing, or horse riding. In recent years, middle-class taste in food in the Romanian capital mirrors global gastronomic and nutritional trends, such as fashionable ethnic cuisines, veganism, norms of healthy eating and eco-conscious consumption, the use of trending spices and superfoods, and specialty coffee. A myriad of restaurateurs are channelling these trends. Street food is one example; it has radically transformed the urban foodscape with the addition of (mainly) American casual foods – burgers, hot dogs, pulled pork, French fries – with a gourmet twist. Such foods have promptly created a category of gourmet street food, counterpartying an older, more localised, popular taste in street foods, such as bagels, pastries, *mici*, and *shawarma*. In keeping with Sharon Zukin's (2008) 'latte towns', food has fuelled the spatialisation of the rising inequality of Romania's capital as new business districts and residential developments are anticipated or joined by middle-class consumption venues, which include world cuisine restaurants, gourmet food trucks, open-air food courts, and artisanal markets. As this emerging cosmopolitan gourmet food offer is becoming more widespread and readily available, a segment of Bucharest's middle class has started to explore outside this globalised class taste into an archive of local tastes.

Foodies as *flâneurs*: contesting taste-based class boundaries

The effacement of taste-based class hierarchies in food choices (Cappellini *et al.* 2015; Guthman 2003; Mellor *et al.* 2010; Roseberry 1996; Simon 2011; Warde *et al.* 1999) paves the way for 'foodies' to emerge as a new category of consumers. Departing from Pierre Bourdieu's theory of distinction, which referred to taste as 'the practical affirmation of an inevitable difference', asserted 'through the refusal of other tastes' (Bourdieu 1984: 56), foodies erode the logic of class-based divisions of taste. They identify through cultural omnivorousness – a move towards cultural eclecticism as a way to affirm social status (Peterson 1997; Warde *et al.* 1999) and through a self-professed intellectual knowledge of food. According to Josée Johnston and Shyon Baumann (2010), foodies distinguish themselves through selective interests in food that are not ranked according to class-bound taste borders, privileging authenticity and exoticism, which they use as criteria to navigate the tensions between democratic inclusivity and status distinctions.

Food engagements 'connote, and by extension, confer status and power on those who know about and enjoy them' (Naccarato and LeBesco 2012: 3) consolidating 'culinary capital'. However, for consumer society, culinary capital is fuelled more by 'an individual's openness to a range of experiences' (Naccarato and LeBesco 2012: 9) than by the exclusiveness of one's taste.

The term is tributary to Bourdieu's cultural capital, designating the collection of tastes, skills, credentials, and more that are acquired by the members of a social class, potentially in the absence of intentionality (Bourdieu 1986). In turn, culinary capital, as suggested by Peter Naccarato and Kathleen LeBesco, is anchored in strategies of self-improvement, and thus more detached from a deterministic class taste. The foodies' struggle for legitimation hinges on preserving the freedom to navigate and explore outside class taste, claiming flexibility and anti-elitism while avoiding serial, standardised products and experiences. This includes selective appreciation of things peripheral, of low status. The logic behind the foodie construction of distinction is not immediately evident to the outsider, as it is based on the skilled mastering of different and diverging codes. In the case of Bucharest's foodies, the target is often food that defies the middle-class culinary restraint and preoccupation with the body, such as the fatty minced meat of *mici* but also tripe soup; offal; or novelties such as lard, bacon, and pig rind obtained from someone's home production. The exploration is not limited to 'out-of-character' types of foods, but, as I will show further, to places 'off the beaten track' in the urban food geography. The range of choices is in itself privileged and a marker of class status.

This choice of satiating, heavy red meats recalls Bourdieu's taste of necessity (1984), associated with working-class food choices, where food is chosen for its function and immediate gratification rather than its aesthetics. Naccarato and LeBesco (2012) identify the 'junk foodie', a middle-class individual entertaining an 'out-of-character' consumption that is anchored in the taste of necessity. This character is in contrast to middle-class food culture where the indulgence in foods and practices is codified as popular taste. By engaging with various 'sites of resistance', such as fairs, carnivals, and competitive eating contests, the consumer takes advantage of 'festive frames' to cultivate temporary hedonic excesses. Such an immersion is consistent with a *flâneur*-like experience of the urban foodscape.

Benjamin (1996) described the *flâneur* as an archetype of modern urban life, acting as a spectator to the urban experience, amateur detective, and decipherer of urban texts: watchful yet reserved, trying to fade away in the crowd. David Frisby (2014) assimilates the *flânerie* envisaged by Benjamin to 'a form of looking, observing (of people, social types, social contexts and constellation), a form of reading the city and its population (its spatial images, its architecture, its human configurations)' (2014: 82–83), from a position of spatial and (middle-) class marginality. The foodies interacting with Bucharest's urban space have the cultural capital (Bourdieu 1984) to act as *flâneurs*: focused on the aesthetics of the city, engaging with street food's sensory cues, they experience the city with a sense of detachment, as spectators to its many facets, seeking to acquire and perform culinary capital. The *flâneuristic* dimension of foodism pushes consumers to seek less-structured consumption spaces and serendipitous, transgressive eating temporalities: bodegas in working-class neighbourhoods, pop-up restaurants, ephemeral open-air venues set up in derelict spaces with fuzzy property regimes, and market food stalls. One review of the *mici* bodegas of Bucharest in *Vice* magazine illustrates such a *flâneur* perspective:

> I love these quaint bodegas, because . . . you get such authentic discussions. Forget conferences of intellectuals debating the fate of humanity in pompous words. . . . After we finished with the *mici*, we lingered around. We made friends with three guys who were socialising after work. This is the charm of these places. Simple, earnest folk.
>
> (Istrate 2016)

This type of exploration is part of a subtle reshaping of the city's geography (from centre to periphery) and temporality (in moments of consumption). The celebration of Labour Day in a peripheral setting such as Obor can arguably summon a temporary carnivalesque setting similar to the sites of resistance frequented by Naccarato and Lebesco's junk foodies: detached from middle-class time and space, permitting anything 'out of character', with no enduring consequences over middle-class selves. Furthermore, the association of *mici* with the street food category lends them an aura of festival food. This frame eases their incorporation into the middle-class foodie consumption portfolio as an experience of hedonic escapism, a suspension of the restrictions associated with the middle-class taste for healthy, 'clean' eating; light foods; and rationality. In the words of an interlocutor: 'In the face of such rapidly changing scientific recommendations, it feels liberating to throw caution to the wind and deep fry a Big Mac – or at least fantasise about doing it'. Some of my interlocutors compartmentalise eating behaviours by granting themselves such temporary opportunities of access to meaty, greasy street food. They perceive these consumption experiences not as nutritional moments, but as temporary, playful, escapist opportunities. This view absolves them of the burden of guilty eating: middle-class consumers become *flâneurs* who venture in and out of the realm of working-class foods and hedonic experiences.

I spent my first May Day at Obor in 2018 with some foodie acquaintances and their friends, who were already familiar with the place: a professionally mixed group of architects, corporate managers, and artists. The chosen bodega, Terasa Obor, was the most visible and prestigious *mici* eatery in Obor, judging by the Facebook posts of the foodies I was following online. It had its own Facebook page that would often post tongue-in-cheek comments and visuals, mostly related to hedonic indulgence. Its imagery often included an aesthetic associated with food porn: close-ups of dripping *mici* or static short films showing masses of *mici* swallowed by smoke as they were being grilled. Due to this increased interest in the iconic meat roll, some downtown places were already experimenting with gourmetised versions. One such gourmet offering included mutton *mici* with horseradish puree instead of mustard, paired with Chandon champagne.

The eatery we were in had become a key site in the urban foodie geography; its patronage was essential for the acquisition of foodie credentials. The group

I was in sat around a shared outdoor table that had taken ages to secure. People took turns standing up. After over an hour of waiting, it took the table only minutes to devour the few dozen hot *mici* off pieces of cardboard splashed with watery mustard. We had eaten so fast and with such anticipation that our heavy stomachs were slowing down the conversation. The spring breeze had swiftly cooled off the food remains.

As people were lighting their cigarettes while contemplating the dried-out greasy meat juices, the crumbled cheap bread, and the trails of oxidising mustard, I mentioned my initial plans to eat gourmetised *mici* at a sophisticated bistro downtown. Someone exclaimed, 'Don't they make things like broccoli soup and posh quiches there? Who would want to go there for *mici*?' People at the table laughed at the prospect, in complicit understanding of how the reference to the two cosmopolitan dishes, the broccoli soup and the quiches, served as metaphorical code for class-adequate taste: light and healthy superfoods and French gastronomy. The suggestion that a venue identified as catering to such a classed taste would try to appropriate *mici* and pass as a popular venue for May Day was considered outrageous. An authentication tournament of the world of *mici* ensued, interspersed with ironic self-aware remarks about the transgressive, gentrifying position we found ourselves in as middle-class intruders at the

Figure 12.1 Table of *mici* at Terasa Obor

Source: Photograph by Monica Stroe

eatery. To drive this point of transgressive consumption conditioned by class privilege home, a fortnight after 1 May, an acquaintance illustrated two of her recent meals in a Facebook post: oysters with champagne at Berlin's Markthalle IX and *mici* with cheap beer at Obor. In the photo caption, tagged 'from Obor to champagne', she reflected on her eclectic lifestyle.

The local foodies' consumer portfolios are enhanced by their privileged knowledge to navigate and immerse themselves in novel settings and to attach self-enhancing meanings to these experiences. They seemingly ignore normative, class-appropriate tastes and eating experiences, glorifying simplicity instead and professing humble appreciation for basic foods and eating moments. By endorsing transgressions as a form of guilty pleasure, middle-class consumers enact a form of symbolic violence over the lower classes. By enjoying the luxury of indulging in cheap, greasy food as an escapist sensorial experience, foodie *flâneurs* profess a so-called simple taste, a distinction strategy typically described by Dana Kaplan as 'omnivorousness': 'a set of aesthetic sensibilities, emotional rules and moral judgments' (2013: 246) whose cultivation enables middle-class consumers to build a classed taste regime based on personal authenticity. This performed cultural capital adds great value for establishing foodie credentials. At the core of these strategies lies the *flâneur*'s urban quest for 'authentic' encounters with 'real' people ('crowds') and 'real' foods.

Consuming (at) the periphery: the spatial organisation of distinction

Labour Day at Obor cultivates several objects of consumption: apart from working-class foods, eating practices, and festive temporalities, social space is likewise consumed. Cultivating a taste of place in urban settings can pave the way for sites to be converted to places of 'eatertainment' (Sassatelli *et al.* 2015), where consumers seek certain food venues for the hedonic experience of an amusement park: a non-consequential exploration outside one's everyday routine.

On May Day in 2019, I was planning to attend a gourmet *mici* event playfully titled *Mici* Saint Laurent, which promised not only *mici*, but also mutton ribs, 'for the more sophisticated'. The title's play on words connected working-class food with fashion icon Yves Saint Laurent. The visuals were also playful: posters with the aesthetics of a glossy fashion magazine cover showed a *mici* roll fitted into a stylish lipstick used by a male model to smear mustard over his lips. Tempting as it was, I had the controversial gentrifying fee-based guided market tour cancelled by Obor purists still fresh in my mind. I thus decided to visit the market instead, to get a feeling of how the various foodie groups were claiming its *mici* stalls.

Upon approaching the market, I am met by an orchestra of smoke vectors that from afar look as if the place was on fire. As I start crossing the open space through alleys of fruit stalls in order to reach the same Terasa Obor of the previous May Day, each smoke vector becomes associated with a terrace selling *mici* grilled on the spot. On arrival, I notice the terrace is shielded by a queue of about 90 people winding itself around the market hall, obstructing the transit of passers-by.

The crowd is an eclectic assortment seemingly dominated by neighbourhood people, most of them in their 50s and 60s, modestly clothed and well wrapped against the strong wind. As I settle in line, I am able to discern a significantly younger, cosmopolitan crowd that stands out through personal styling, clothing design, and accessories: oversize glasses, piercings and tote bags; fashionable sneaker brands; asymmetrical haircuts; some daring hair colours for women; and thick, trimmed beards for men. Occasionally, group selfies are taken, paired with self-referential comments and giggles about the awkwardness of queuing for food. They look like well-defined bubbles, interspersed through the more local-seeming, older crowd. I can recognise many familiar faces from the intersecting constellations of the creative and academic middle classes: a group of film and theatre professionals, the members of a left-wing LGBT organisation, a wine sommelier with her friends. There is also the group of a *Vice* magazine journalist engaged in a *mici* stall crawl through the city's peripheral venues to document a gastronomic review of low-key *mici* stalls. Newcomers keep joining their friends in the queue, thickening the line ahead. 'I really hope this place

Figure 12.2 Crowds queuing at Terasa Obor

Source: Photograph by Monica Stroe

you brought me to is going to prove worthwhile!' exclaims a latecomer girl in her 20s to one of the *flâneur* groups, concerned about the foreseen waiting times. The shared experience of waiting helps homogenise the two crowds into a vague atmosphere of boredom combined with anticipation; they exchange *mici*-related knowledge. I overhear a middle-aged man advising in an expert tone some sneaker-wearing youth that, in order to receive the optimal quality under conditions of rushed orders, one needs to specifically request *mici* from the third row of the grill as they are not charred but not undercooked either.

The limited seating means that *flâneurs* often share tables with the house regulars. A couple of friends, acquaintances, and friends of friends join my table. As our number grows, we also eagerly join a larger table, where two older men are lingering over draft beer. The physical intimacy creates ambivalent situations: the crowds of regulars spatially integrate the *flâneurs* but simultaneously apprehend them visually as intruders and eavesdrop on their conversations like anthropologists during fieldwork.

In the proximity, I recognise artists and academics who had gathered via a Facebook event I was following online. In an obvious throwback to the socialist-era May Day celebrations, the online event initiator had urged the participants to come dressed proletarian and to know the lyrics of the socialist regime's May Day official workers' song, '*1 Mai Muncitoresc*' ('First of May, Workers' Day'). To frame his event, the organiser used the ironic slogan '*Mici* of the world, unite!' Appropriating stereotypical working-class symbols and reworking them into the events of foodie *flâneurs* is a recurrence visible in other Labour Day advertisements. Telling slogans abound: 'New Year's Eve for the working man', '*mici* to the people', *mici* as 'the working-class burger'. As in a tournament for exclusivity, an outlier member, a female visual artist, challenges the group's enthusiasm on the Facebook event page. She had manifestly boycotted the gathering in favour of an even more off-the-grid *mici* stall in the same market that, as she phrased it, 'was thankfully lacking Puma- and Converse-wearing hipsters' who had, allegedly, already 'ruined' her favourite tavern. Such a statement illustrates foodies' engagement with popular tastes as a distinction strategy and their claim to space as a means to secure the symbolic monopoly over authenticity and exoticism.

Development and urban regeneration processes are already set in motion in the area, with a controversial shopping mall inaugurated near Obor Market in 2016. While the shopping centre was heavily criticised and boycotted by the market's *flâneur* enthusiasts, their eatertainment trips to the marketplace are also expected to contribute to the acceleration of such processes of displacement and exclusion. Referring to popular consumption venues, such as ethnic groceries and diners, Zukin argues that the gentrifiers' nostalgia for these quaint social spaces of 'authentic' neighbourhoods engages them 'in a voyage of discovery' (2008: 735–736). Authenticity, then, 'becomes an effective means for new residents to cleanse and claim space' (2008: 747). These consumption venues magnetise foodies' culturally omnivorous taste for both 'high' and 'low' foods and food practices (Johnston and Baumann 2010). Foodies claim and potentially shape the Obor market space based on an ambivalent stance. On the one hand, they try to position their transgression as a legitimate exercise of distinction through culinary capital. On the other hand, they also strive to transgress mainstream practices as a manifest for the democratisation of taste: entertaining an appearance of value-free, indiscriminate consumption. The spatial implications of foodies' omnivorous taste recall Zukin's observation about the gentrifying effect of consumers' hybrid subjectivities over American cities: '[T]his [hybrid] subjectivity commands the high culture of a fine Bordeaux wine as well as the low culture of a kosher pickle' (2008: 735).

Behind the curtain of smoke: affective atmospheres, qualisigns, and the senses

The *flâneurs*' enacted appreciation of sensorial elements pertaining to the cheap minced meat rolls associated with the affective atmosphere elements of their consumption is based on the reshuffling of the qualia or qualisigns (Harkness 2015; Meneley 2008; Weiss 2012), accessible via mediation from class taste. Qualia are 'pragmatic signals that materialise phenomenally in human activity as sensuous qualities' (Harkness 2015: 573). In the realm of qualisigns, the Obor market *flâneurs* articulate pedagogies pertaining to culinary capital such as the optimal texture of *mici* (loose and juicy), a desirable cooking time (with middle-class taste privileging those that are medium done rather than charred), as well as enthusiasm for the overpowering garlic content. Such taste is enacted discursively in some of the gastronomic reviews of *mici*, which sometimes underline sensorial dimensions with terms specific to fine dining reviews: 'crust', 'charred smell', 'notes of garlic', and 'well-balanced flavours' (Costache 2019).

Affective atmospheres as modes of spatialising affect (Bissell 2010) frame the *flâneur* experience as foodie visitors are guided through the *mici* stalls experiencing a 'distributed yet palpable . . . quality of environmental immersion that registers in and through sensing bodies whilst also remaining diffuse, in the air, ethereal' (McCormack 2008: 413). Ben Anderson envisages affective atmospheres as fluid assemblages of objects, discourses, and people in everyday situations, 'affecting one another as some form of "envelopment" is produced' (2009: 3). The 'envelopment' of celebrating bodies is at play on the socially eclectic site of the Obor eatery, as the ethnographic notes suggest. At the May Day Obor table I am at, the conversation develops with debates and demonstrations of the proper way to pick up the meat roll with the fork-replacing toothpicks. One person points to a local patron to illustrate eating skills, pointing out the need to hold the roll with not one but two toothpicks in order to balance it. Two people take phone snapshots of soiled *mici* cardboards and grease leftovers to be posted online, commenting together about the frame's aesthetics. At times foodies acknowledge and evaluate the music pieces from the speakers as either fitting with the atmosphere of the place or – when too much in tune with middle-class taste – regard them as unpleasant, out-of-place auditory presences. In a seemingly symbolic tournament of thrill-seeking and expertise display, people at the table share news from social network newsfeeds about what other novel outposts of *mici* other friends have visited in the meantime, as well as recollections of memorable past experiences and firm opinions about what makes the perfect *mici*. Pinning down the visited places in social media posts represents a key dimension of the explorations, as this virtual flagging of exclusive places contributes to the build-up of the *flâneur* persona. It is also a form of classed place-making, and, aggregated, it creates an online map of the outposts of the foodie siege of the periphery.

I experience Labour Day at Obor with the thrill of a guilty pleasure that arises from the immersive effect of sensorial cues: the overwhelming, eye-watering smoke vectors; music; buzzing crowds; the sizzling of the grill; the strong fried food smells stigmatised as working class; the fat minced meat of *mici* that drips from my fingers with no forks to handle the *mici*; and the strong garlic taste in my mouth. As a middle-class newcomer, I am required to do emotional work to become receptive and to engage with the place's 'vibe'. Aggregated, the overpowering sensorial stimuli summon the immersion into an affective atmosphere of the marketplace that challenges the everyday comfort of the middle-class eater. Eavesdropping and acquisition of skills from local patrons entertain the illusion of being part of a collaborative process of co-creation of place. They, in fact, obstruct the appropriation processes at play in the marketplace.

Conclusion

This chapter follows the emergence of Bucharest foodie *flâneurs* on the background of a middle-class urban foodscape, by now saturated with burger trucks and corporate-sponsored festivals, which have succeeded in legitimating street food as adequate to middle-class taste but quickly limited foodies' access to meaningful, non-standardised, and non-controlled street food experiences. Avoided and despised for decades, *mici* have benefitted from the explosion of this cosmopolitan street food market to become a foodie alternative to staged, contained street food experiences. The foodies seek out the sensorial dimension of these explorations as a form of liberation from middle-class bodily restraints and nutritional principles.

In the current globalised consumer universe of Bucharest, eating *mici* and wandering to the urban periphery represent the *flâneurs*' attempts to identify experiences that provoke the senses and to expand their consumer portfolios. I have identified the foodie *flâneur* as an urban character engaged in a spatial-temporal exploration of carnivalesque, event-based consumption of peripheral street food enacted at a peripheral location as a performative exercise in culinary capital. The appropriation includes both liminal spaces (the chaotic, barely regulated marketplace) and working-class consumption. Decades after the irreversible demise of the symbolic capital of the working-class with the fall of state socialism, a segment of the postsocialist middle classes sets out to re-signify such symbolic resources (Labour Day, working-class foods) by appropriating them as performative exercises of cultural omnivorism. The 'hybrid subjectivity' of urban gentrifiers, discussed by Zukin (2008), fuels, in the case of *flâneur* foodies, a hybridised repertoire of both high and low food taste.

Flâneurs go in and out of these spaces, without long-term commitment, in search of novel escapist, hedonic experiences of playing working class. However, they leave their aestheticising mark on the venues that they frequent as business owners start acknowledging them as valuable customers and start catering for the projected *flâneur* taste: upgrades in terrace furniture, witty ads, and new music repertoires. There are also online traces of this acknowledgement, such as the initiative of businesses to create social media channels and to upload enticing, Instagram-styled food close-ups. The *flâneurs'* online marks of their explorations – hashtags, grocery-shopping posts, photographs of peasant grannies or mile-long *mici* queues, the transcription of quaint conversations overheard at the terrace – build up an attractive image of these temporary hotspots. These transgressive consumption trips make a symbolic but likely long-lasting contribution to the changes to the postsocialist urban space to gentrification and displacement, which add to growing pressure from real estate and other capitalist forces on working-class neighbourhoods and peripheral areas.

References

Anderson B. (2009) 'Affective Atmospheres', *Emotion, Space and Society* 2 (2): 77–81.

Benjamin W. (1996) *Selected Writings: 1938–1940* (Vol. 1–4). Cambridge, MA: Harvard University Press.

Berdahl D. (2005) 'The Spirit of Capitalism and the Boundaries of Citizenship in Post-Wall Germany', *Comparative Studies in Society and History* 47 (2): 235–251.

Bissell D. (2010) 'Passenger Mobilities: Affective Atmospheres and the Sociality of Public Transport', *Society and Space* 28 (2): 270–289.

Bourdieu P. (1984) *Distinction*. Cambridge, MA: Harvard University Press.

Bourdieu P. (1986) 'The Forms of Capital', in Richardson J. G. (ed) *Handbook of Theory and Research for the Sociology of Education*. New York: Greenwood Press: 241–258.

Cappellini B., Parsons E. and Harman V. (2015) '"Right Taste, Wrong Place": Local Food Cultures (Dis)identification and the Formation of Classed Identity', *Sociology* 50 (6): 1089–1105.

Costache D. (2019) 'Am colindat Bucureştiul de 1 Mai ca să văd cine e regele micilor (eu şi alţi nouă experţi)', *Vice*, 2 May, www.vice.com/ro/article/evyxjn/regele-micilor-din-bucuresti.

Frisby D. (2014) 'The Flâneur in Social Theory', in Tester K. (ed) *The Flâneur (RLE Social Theory)*. London: Routledge: 81–110.

Guthman J. (2003) 'Fast Food/Organic Food: Reflexive Tastes and the Making of "Yuppie Chow"', *Social & Cultural Geography* 4 (1): 45–58.

Harkness N. (2015) 'The Pragmatics of Qualia in Practice', *Annual Review of Anthropology* 44: 573–589.

Istrate M. (2016) 'Am colindat terasele cu mici din Bucureşti, ca să văd care sunt cei mai buni', *Vice*, 20 August, www.vice.com/ro/article/8qnk44/am-gustat-micii-din-bucuresti.

Johnston J. and Baumann S. (2010) *Foodies: Democracy and Distinction in the Gourmet Foodscape*. New York: Routledge.

Kaplan D. (2013) 'Food and Class Distinction at Israeli Weddings: New Middle Class Omnivores and the "Simple Taste"', *Food, Culture & Society* 16 (2): 245–264.

Kideckel D. (2010) *România postsocialistă. Munca, trupul și cultural clasei muncitoare*. Iaşi: Polirom.

McCormack D. P. (2008) 'Engineering Affective Atmospheres on the Moving Geographies of the 1897 Andrée Expedition', *Cultural Geographies* 15 (4): 413–430.

Mellor J., Blake M. and Crane L. (2010) '"When I'm Doing a Dinner Party I Don't Go for the Tesco Cheeses" Gendered Class Distinctions, Friendship and Home Entertaining', *Food, Culture & Society* 13 (1): 115–134.

Meneley A. (2008) 'Oleo-Signs and Quali-Signs: The Qualities of Olive Oil', *Ethnos* 73 (3): 303–326.

Naccarato P. and LeBesco K. (2012) *Culinary Capital*. London and New York: Berg.

Peterson R. A. (1997) 'The Rise and Fall of Highbrow Snobbery as a Status Marker', *Poetics* 25 (2–3): 75–92.

Roseberry W. (1996) 'The Rise of Yuppie Coffees and the Reimagination of Class in the United States', *American Anthropologist* 98 (4): 762–775.

Sassatelli R., Santoro M. and Semi G. (2015) *Fronteggiare la crisi: come cambia lo stile di vita del ceto medio*. Bologna: Il Mulino.

Simon B. (2011) 'Not Going to Starbucks: Boycotts and the Out-Scouring of Politics in the Branded World', *Journal of Consumer Culture* 11 (2): 145–167.

Warde A., Martens L. and Olsen W. (1999) 'Consumption and the Problem of Variety: Cultural Omnivorousness, Social Distinction and Dining Out', *Sociology* 33 (1): 105–127.

Weiss B. (2012) 'Configuring the Authentic Value of Real Food: Farm-to-Fork, Snout-to-Tail, and Local Food Movements', *American Ethnologist* 39 (3): 614–626.

Zukin S. (2008) 'Consuming Authenticity: From Outposts of Difference to Means of Exclusion', *Cultural Studies* 22 (5): 724–748.

13 Michelin stars and *pintxo* bars in Donostia

Taste, touch, and food tourism in contemporary urban Basque Country

Aitzpea Leizaola

A substantial part of the Basque culture revolves around food, gastronomy, and conviviality. A major local and national identity marker, food plays an important role in contemporary Basque Country. Since the 1980s, Donostia, the capital city of Gipuzkoa in the Basque Autonomous Community on the Spanish side of the border, has become the centre of a vivid and dynamic gastronomic culture and, as a result, a main tourist attraction. This chapter focuses on the impact tourism has drawn with respect to *pintxo* eating – a particular commensality practice that originated in Donostia. Highly praised locally, *pintxo* eating appeals to a sensual approach to food in an urban sociability. In recent decades, tourism has spurred significant changes in *pintxo* eating that are not always well received locally.

Mentioned several times in the travel supplement of the *New York Times*, Donostia was described as a laid-back surfing resort that was 'not for weak-willed dieters; there are sweet-smelling cafes and mouth-watering bites to tempt you at every turn' (Lee 2007). For once, the beautiful seafront that had been the pride of locals and the joy of visitors for over a century had been left aside. Instead, the focus has become gastronomy.

Since the nineteenth century, the Basque Country has been a major tourist destination for European aristocracy and the 'well-to-do'. Following the European trend at the time, the Atlantic coast attracted both the political and economic elites as early as the 1840s. While French emperor Napoleon III favoured Biarritz, Donostia became a seaside resort preferred by the Spanish royals who moved there for the summer. The proximity of the border played a key role in its tourist development since the end of the nineteenth century and during the turbulent beginning of the twentieth century up to the outbreak of the Spanish Civil War in 1936. Throughout his dictatorship and up until his death in 1975, Franco also set his summer residence in Donostia. For over a century, the city was the vacation destination of the rulers. This deeply shaped the city's development for subsequent decades, no doubt conditioning the type of tourist who could afford to visit.

Today, Donostia's heritage, its natural environment, its effervescent cultural scene, including events such as the San Sebastian International Film Festival and Jazzaldia, together with traditional festivities and celebrations, constitute well-established attractions. However, the recent tourist boom owes much to both Basque food culture and the place gastronomy has reached on a global scale. With internationally renowned restaurants and chefs, Donostia is now a top international destination for food tourism, with the Basque Country recurrently appearing in specialised blogs and high-impact media posts, such as the HuffPost and *National Geographic* (Atxa 2016; Satran 2014).

Every year hundreds of thousands of tourists[1] eagerly flock to taste the delicacies of Basque cuisine that include outstanding seasonal produce – a large variety of crops, but more substantially mushrooms, game, and above all, fish. Essentially defined as a 'market cuisine', Basque cuisine today encompasses both the most sophisticated high cuisine and more traditional cooking of large portions of homemade dishes, the *sagardotegiak* (cider places) or the countless bars that prepare delicacies of miniature cuisine. Recently, tourist-oriented gourmet shops – a phenomenon long noted in the Basque provinces on French territory – have opened their doors in Hegoalde on the Spanish side, too, displaying a thriving offer of terroir products. Indeed, since the foundation of the Basque Autonomous Community in 1979, the Basque government has mentioned local gastronomy in its campaigns for tourism promotion. Taste now occupies a major place in the construction of a sense-orientated, place-based experience for visitors and dwellers alike.

The changes within the *pintxo* culture are a good example of the deep transformations occurring in the Basque urban gastronomic scene. Considered a culinary art piece on its own for the most sophisticated ones, the *pintxo* (literally 'spike' or 'skewer') is a small, one-bite appetiser consumed in bars, generally when going out with friends (see Figure 13.1).

Its name refers to the toothpick holding together the food on top of the bread slice. For many, it is an essential part of contemporary Basque cuisine as well as one of the best ways to taste the latest creations in culinary innovation for a reasonable price. Going for *pintxos* is one of the most popular things locals do when hosting visiting friends. Not surprisingly, it has become one of the main experiences advertised by the city's tourism office and by tourist enterprises.

Part of the data presented in this chapter is based on a research project on the evolution of Basque gastronomy at the eve of the millennium funded by the Basque government from 2009 to 2010. Because of the centrality of Donostia in the creation and spread of the new Basque cuisine, a significant part of the fieldwork – over 40 interviews with bartenders, caterers, chefs, and restaurants owners, as well as several discussion groups gathering dwellers from different social backgrounds – was conducted in the city. Since 2009, I have continued long-term ethnography, during which *pintxo*-eating practices were studied through informal conversations with tourists and dwellers. The monitoring of various 'foodie' blogs provided further insights into how *pintxo* eating is perceived as a 'must-do' for visitors to Donostia.

In recent years, a growing emphasis has been noted on studies on tourism experiences related to food (Mak *et al.* 2012). Since early 2000, gastronomy

Figure 13.1 Pintxo eating

Source: Photograph by Aitzpea Leizaola

has drawn considerable attention as a tourist product, even if the relationship between tourism and gastronomy remains to a certain extent peripheral (Medina *et al.* 2018). Indeed, the phenomenon has received different names in academic writing, as 'food', 'culinary', and 'gastronomic' tourism. Food tourism strategies are a sound instrument of regional development due to their potential leverage between agricultural production and tourism (Hall and Sharples 2003). But not only. In the Basque Country, as well as in other parts of the world, entrepreneurs and politicians have noted the potential of food tourism to attract first-time visitors to encourage them to come back for 'seconds'. Today, the tourist has become – and is thought of as – a consumer, not only of goods and services but, above all, of experiences. Unlike mass tourism, much contemporary tourism focuses on authenticity and uniqueness. Gastronomy is a core part of this experiential focus, where food is presented and perceived as a conveyor of unique sensory experiences.

Over the past twenty years, both the anthropology of food and the senses have gained substantial interest by researchers. A substantial body of literature explores food as a source and marker of social distinction, but relatively few authors analyse the ways that the senses play into these processes (Sutton 2010). However, in both tourism and advertising in general, food appears as a major sensorial trigger, an enticing item capable of influencing people's choices, be it in terms of holiday destinations or product consumption. The sensory aspects of food appear prominently connected to notions of place and identity. In this perspective, the attachment of taste to place can be seen as one of the tautologies of food and identity (Sutton 2010: 216). Taste, image, origin, experience, and quality are now all recognised as important 'not only because of the role of food in the local economy but also because what, why and how we eat says something about ourselves, why we travel and the society we live in' (Hall and Sharples 2003: 2). As Erik Wolf (2020), executive director of the World Food Tourism Association puts it, 'food tourism is the act of traveling for a taste of place in order to get a sense of place'.

Donostia's recent transition well illustrates the connection between food tourism, senses, and the production of place. All five senses are involved in *pintxo* eating. Whereas touch and hearing render primarily to the social aspect of eating and are particularly relevant in the crowded and noisy *pintxo* bars, as in any eating experience, sight, smell, and especially taste are deeply involved when going for *pintxos*. Eating with the eyes comes first: besides the counters full of tempting food bites, placed inside and outside bars, pictures displaying different bites and dishes are intended to guide tourists on what to order. But above all, taste is central. *Pintxo* eating is definitely a sensorial experience. And there is nowhere better to experience it than in the tiny bars of Donostia's Old Town where it all started.

Donostia, the jewel of the Basque cuisine

The Basque Country is a major gastronomy destination, and tourists come to enjoy the much-praised Basque cuisine. For many, Donostia represents a mandatory stop. As multi-starred chef Andoni Luis Aduriz from the world-famous restaurant Mugaritz acknowledged, a large part of their customers are foreigners who travel specifically in search of new culinary realities. For them, the city is worth a plane ride. This is not surprising, since his restaurant has been listed several times as the fourth best restaurant in the world. With 34 Michelin stars distributed in 24 restaurants in the Basque Autonomous Community as of 2020, the Basque Country is sixth in the 'Top 10 Best Culinary Destinations in the

World', where no less than sixteen are located in Donostia and its surrounding area (Tourist Maker n.d.). Apart from Kyoto in Japan, no place in the world displays such a high concentration of Michelin stars per square metre, as the tourism office website proudly boasts (DSST 2020). Indeed, Donostia is listed second in the twenty most-starred Michelin cities in the world (Satran 2014). Other prestigious rankings ratify the French distinction: in 2019 five restaurants counted among the '50 Best Restaurants in the World', a ranking in which Basque restaurants have occupied leading positions continuously since 2006 (TW50B 2020). As these awards show, Basque chefs have obtained worldwide recognition, representing a main attraction for food tourism. This transformation took over four decades to transpire. In 1980 eleven young chefs founded the New Basque Cuisine, triggering a real revolution in Basque gastronomy. For some, their impact goes well beyond the tiny Basque Country to mark 'the origin of modernity of Spanish cuisine' (Fernandes 2009; González and Corcuera 2008). Donostia is the centre – and, indeed, driver – of this process.

The phenomenon of high-end restaurants goes well beyond an economic issue – it also permeates issues of class and distinction, where experienced gourmet 'foodie' travellers and the average holiday makers may meet as they head to Donostia's *pintxo* bars in the Old Town in search of an 'authentic', singular experience. At stake is not only food itself or the feeling of exclusivity of high-end restaurants, but rather the possibility of 'going local' through an essentially sensorial experience. *Pintxo* eating gives the tourist a feeling of being fully immersed in a local experience while tasting the latest in high cuisine at an affordable price.

The *pintxo*, a deeply rooted invented tradition

Made of *piperrak* (Basque pickled, long, tiny chillies), olives, and anchovies skewered on a cocktail toothpick, the *gilda* summons the quintessence of the *pintxo*: customers pick it up with their fingers from the counter and eat it in just two mouth bites, being careful not to get oil stains on their clothes. The art of preparing *gildas* goes hand in hand with the art of eating them. Named after Rita Hayworth's character Gilda in the eponymous film directed by Charles Vidor that was released in 1948 at the San Sebastian International Film Festival, this popular appetiser is considered to be the first *pintxo*. Its origin goes back to Casa Vallés, a wine tavern run by two Navarrese brothers in the centre of Donostia (Azpeitia 2016). 'Green, salty and hot'– where, in Spanish, the adjective 'green' is colloquially used to mean lustful, while 'salty' has a clear sexual connotation when referred to women – this *pintxo* is lustful, exuberant, and tempting, just like Gilda (González and Corcuera 2008).

The *gilda* was a success among the *txikiteros*, groups of male-only friends that would go from bar to bar drinking small glasses of cheap wine and paying rounds in turns. *Txikiteo* (from *txikito*, 'small' in Basque) refers to the smallest wine glass that is served in bars where this activity became a major pastime for men after the war, carried out on a daily basis, before lunch and then again in the evening after work. The history of the *pintxo* goes hand in hand with that character. Its name played very much on gendered references and, as a result, became an edible metaphor.

An incarnation of the very essence of the femme fatale, Gilda – the popular sex symbol – was the very antipode to the pious girls and demure housewives who embodied the feminine ideal promoted by Franco's dictatorship. This *pintxo* conveyed highly sexual components, both in terms of the act of eating itself, which has intrinsically erotic connotations, and of the *gilda* in particular, enhanced by the fact of taking the food to the mouth with one's own fingers. In a society with strict sexual and gender divisions concerning sociability and leisure, especially in urban contexts, and under the direct control of the Catholic Church, as was the Basque Country during Franco's dictatorship, desire could not be publicly displayed. The control of sexuality and sexual references in general was one of the pillars of the regime, upheld through censorship. As older informants recall, when the film was released in Spain, it was believed that the scene in which Gilda takes off her glove anticipating a striptease had been previously cut by Spanish censorship. Nowadays, few are familiar with the origin of the name, and *gilda* has become a common noun in Spanish to refer to skewered pickles on a toothpick.

The success of the *gilda* was quickly followed by the creation of new *pintxos* or *banderillas* ('bullfighting spikes'), as they were called at the time. Unlike tapas, usually offered by the bartender for free, consumers always need to pay for *pintxos*. *Pintxos* are a good example of a successful invented tradition. From the 1950s, as more bars prepared different specialties, *pintxos* became a specificity of Donostia (González and Corcuera 2008). As a result, they were incorporated into the city's atmosphere of conviviality, and with time they have become both typical and a traditional practice of Basque gastronomy. Their success is comprehensible, as *pintxo* eating embodies the conjunction of sensory pleasure and leisure in a conviviality context. Whereas major changes have occurred both in terms of food and in the way they are eaten, as with other invented traditions (Hobsbawm and Ranger 1983), *pintxos* are presented and thought of as traditional, unchanged practices.

Pintxo eating calls to the sensory dimensions of food as part of the construction of a sense of place, which, as we have seen, plays a significant role in Donostia. Since the apparition of the New Basque Cuisine, cooking techniques, stylish plating, and specific ingredients that make up the *savoir-faire* of high cuisine have been popularised through *pintxos*. In this process, high cuisine textures, flavours, and tastes have become affordable for a broader spectre of people, blurring the distinction between the taste of necessity and the taste of luxury defined by Pierre Bourdieu (1984). Some of the most popular first *pintxos* were simple dishes in terms of preparation and cooking techniques, such as deep-frying after dipping the food pieces in egg and coating them in bread crumbs. Similarly, humble animal parts for offal-based *pintxos* were quite popular: kidney, lamb sweetbreads, fried calf brain, pig ear, and calf snouts.

However, even if they are viewed as an intrinsically 'rooted' phenomenon, *pintxos* are not devoid of external influences either. Gastronomy, as well as food practices, evolves, as does taste. The effect of tourism on food practices is not new (Cohen and Avieli 2004). In the Basque context, specific products are prepared to satisfy tourists' expectations, as exemplified by paella and sangria, which are considered to be typically Spanish (Leizaola 2006). In the Old Town as in other parts of the Basque Country, especially in the borderland, only tourists order from the jug of sangria sitting by the beer tap in many bars. Similarly, restaurateurs have introduced or adopted some practices, such as early supper and dinner times, aimed at international tourists.

Taking the plate out

As the 2011 Spanish anti-tobacco law brought customers to the streets, bars took semi-mobile furniture out of their premises: barrels, high tables, stools, and even shelves, as well as semi-permanent terrace structures providing shelter from the rain. This ruling and its subsequent reshuffling of space had a direct impact in the pedestrianised Old Town, making it difficult to walk in its narrow streets, especially in the summer and on other significant holidays, when hundreds of people, dwellers and tourists alike, eat and drink in the street. In the last decade, bars have implemented minor reforms, too, to adapt their offerings – mostly concerning food consumption patterns – to tourist expectations. Some changes were introduced by tourists to be later adopted by bartenders, as was the case of what was locally known as the 'plate issue'.

During my fieldwork, choosing more than one *pintxo* on a plate as if at a buffet was strongly criticised by locals (see Figure 13.2). Indeed, it has become a marker of those who know the rules, generally identified as 'locals', and those who do not. A woman in her 70s, born in the Old Town, reported to me the following incident when going for a drink after attending a funeral last winter:

> We went to the V [a well-known *pintxo* bar], and ordered the drinks in Basque but had to repeat in Spanish, as the waiter did not understand us, even if everything was written in four languages, including Basque. We were having a look at the *pintxos* when he handed us a plate. 'Don't you dare give me a plate!' I told him. 'We are from here!' The waiter, who himself was a foreigner as revealed by his accent, answered, 'Well, madam, it is like this in all the bars of the Old Town'. 'Well', I replied, 'I will not come to these bars, then'.

Handing over or asking for a plate is not a simple gesture. The woman was outraged at having been mistaken for a *giri* (a derogatory term used by locals to refer to tourists[2]). Throughout my fieldwork, I have witnessed incidents of this type in several bars, with levels of confrontation of varying intensity in which local customers were involved: from more or less heated discussions with the

Figure 13.2 Tourist holding a plate full of *pintxos*

Source: Photograph by Aitzpea Leizaola

waiter, as in the previously mentioned case, to refusals to take the plate. Holding a plate entails people circulating back and forth around the counter while filling it up, which deeply clashes with the local way of *pintxo* eating: that is, trying just one *pintxo* per bar before moving to the next one. The plate is sensed by locals as extremely annoying as it defies 'the proper way to do things': that is, elbowing your way to the counter if needed, securing your spot, ordering drinks and hot *pintxos* or choosing cold ones with your fingers, and eating them right away while standing up. Whereas physical contact is unavoidable when the bar is packed, locals tend to keep it as quick and short as possible, as when handling money to pay the bill over other customers' shoulders or squeezing into the counter to get drinks or hot *pintxos* as the bartender loudly shouts the customer's name.

As a woman working in the catering industry for almost 30 years commented to me, the plate issue was a shock for her from the beginning: 'I have always wondered how come nobody tells them [the tourists] at the tourist office that this is not the thing to do!' The plate issue is seen as a sign identifying both *giris* and the bars that target them, accordingly called *giri-bars*. Some years ago, local customers would even leave the bar if they were handed a plate. Nowadays, many locals systematically avoid going into these places, as reported in several interviews. The plate issue is a visible indicator of one of the most perceptible changes that tourism has brought to the Old Town: that is, the change in *pintxo* commensality and its impact in urban space management (see Figure 13.3).

Whereas warm *pintxos*, cooked to order, are always served on small plates, customers would help themselves to cold ones. Twenty years ago, caterers were reluctant to hand over any plates or cutlery for cold *pintxos*. It was common to see bartenders mimicking to foreign customers to help themselves directly from the dishes on the counter. *Pintxo* bars competed with each other to display an ever-filled counter where you could barely leave your glass, much less a standard, shallow plate (see Figure 13.1).

Here, it is interesting to note the ambiguous status of a key feature of the *pintxo*: that is, helping yourself and eating with your fingers, with no plate or

Figure 13.3 Tourists enjoying their *pintxos* out in the street

Source: Photograph by Aitzpea Leizaola

cutlery – perhaps the element that most repels Westerner tourists (Cohen and Avieli 2004). Utensils are not merely hygienic but are also gadgets mediating between the food and the eater, so they are deeply symbolic. Eating with one's fingers epitomises an animalistic behaviour of devouring of food by 'uncivilised' people (Elias 1979). Hence, the plate issue emphasises and mobilises deeper significations. Even if *pintxo* eating may not happen in a particularly exotic environment, but in the rather familiar setting of a bar, for many it nevertheless requires confronting their own prejudices and manners.

Will there be enough people to eat all these *pintxos*?

Coinciding with the end of political violence in the Basque Country and the closure of many popular tourist destinations to Westerners due to fear of Islamic attacks, tourism significantly increased in the Basque Country. At the end of 2019, tourism was no longer considered a seasonal activity in Donostia. Micro and macro political issues, together with the combined efforts of tourism agencies – both local and regional – explain the increase in the number of visitors in the last years.

An extended tourism season, together with a rapid rise in visitor numbers, especially day-trippers, has placed tourism at the top of the urban debate agenda in Donostia (García-Hernández *et al.* 2017). As Manuel Delgado (2011) points out, the heritage order of a place is not immune to the renovation, regeneration, or restructuration that the neoliberal city's transformation entails, nor to the standardisation of behaviours that usually accompanies them. In the case of Basque gastronomy, and more precisely in the context of *pintxos*, this appears clearly in the intricate legal regulations of the occupation of the public space, particularly in the Old Town, and the resistance of the dwellers to the touristification of the neighbourhood. More importantly, it is perceptible, too, in the homogenisation of bars and restaurants, both in style and in terms of food offer.

Often served on specifically designed tableware and demanding cutlery, hot *pintxos* have paradoxically become increasingly sophisticated. They can barely be eaten with the fingers. As plates are systematically handed over to every customer, eating with your fingers is being presented as a challenging experience by several tourism companies to tourists eager to go off the beaten track. Nowadays, several bars in the Old Town have operated a clear shift from drinking to eating. Tourist-oriented bars and the plate system have put a stress on food, often offering simple and not-so-elaborate *pintxos* intended for not-so-well-informed or taste-trained tourists. Whereas tourist-oriented bars focus mainly on cold *pintxos* displayed on the counter, high-end *pintxo* bars specialise in hot, miniature cuisine-style *pintxos*. Overall, the prices of *pintxos* have increased, as foreign tourists generally have greater purchasing power, while their quality has substantially decreased. In some of these bars, they are not even prepared in the bar but brought from outside. As *pintxos* are locally viewed as part of the Basque culinary heritage, a specificity the inhabitants of Donostia are particularly proud of, a strong opposition to such changes has arisen in the last years.

Conclusion

Nowadays, *pintxo* bars are a substantial part of Donostia's foodscape and one of its main assets for tourism promotions. Going for *pintxos* has become a massive tourist phenomenon, around which specific dynamics have been created, such as guided *pintxo* tours offered by various enterprises. Studying particular food practices through the lens of the senses provides insights into how the city and its dwellers as well as its visitors adapt and co-produce each other. In the case of *pintxo* eating, the impact of tourism in Donostia, and especially food tourism, is highly perceptible. In the last two decades, the Old Town has significantly muted; the transformations have been particularly visible in the catering industry, especially the inner organisation of bars and eateries, as well as their food offers, are increasingly tourist orientated. In the last decade, *pintxo* offers have become more homogeneous: old-style *pintxos* like fried hake or cod and overall, offal-based *pintxos* have practically disappeared while some others, like mini hamburgers, are widely prepared.

The plate issue has also had unsuspected consequences. Some of the most reputed *pintxo* bars have recently incorporated a set *pintxo* menu in their gastronomic offer. Halfway between the miniature cuisine and the *giri* way described earlier, these menus represent a significant change in the way of eating *pintxos*. First, as in the plate issue, the set menu, which even includes a dessert, means abandoning *txikiteo*: that is, going from one bar to another, as commonly associated with going for *pintxos*. It also entails spending more money in a go in a single place. On the other hand, it almost systematically requires leaving aside eating while standing up at the counter. It is interesting to note that in contrast to the plate issue, these *pintxo* menus have not drawn much criticism so far among regular local customers.

Counters full of plates with amazing assortments of *pintxos* appeal to tourists and dwellers alike from the streets. However, as we have seen, the changes introduced by tourists, particularly with the piling up of *pintxos* on a plate, deeply upset the locals. Whereas traditional *pintxo* commensality is related to a particular way of eating – standing up in crowded, busy bars, holding the food with one's own fingers – in the new scheme, the whole sensorial dimension of *pintxo* eating has been deeply transformed. In today's tourist-oriented Old Town, taste has been profoundly homogenised, particularly in the case of the *giri*-bars, and the tactile dimension acts no longer as a direct intermediary between food and the individual. While tourists are usually eager to 'go local', often much of their *pintxo* eating experience has occurred on a perfectly orchestrated, tourist-orientated stage.

Notes

1 *Tourist* will be used to refer to those classified as such according to the WTO namely, travellers spending the night on the spot, as well as day-trippers. Indeed, many of the tourists in Donostia are French day-trippers.

2 Pronounced *guiri* in Spanish, this term was shortened from Basque in the nineteenth-century Carlist wars, and today is used as a colloquial, derogatory term for *tourist* in Spanish.

References

Atxa E. (2016) 'Food: Basque Country, National Geographic', www.nationalgeographic.co.uk/travel/2016/04/food-basque-country.

Azpeitia J. (2016) *Pintxos de leyenda de Donostia*. Donostia: Ttarttalo.

Bourdieu P. (1984) *Distinction: A Social Critique of the Judgement of Taste*. Cambridge, MA: Harvard University Press.

Cohen E. and Avieli N. (2004) 'Food Tourism. Attraction and Impediment', *Annals of Tourism Research* 31 (4): 755–778.

Delgado M. (2011) *El espacio público como ideología*. Madrid: La Catarata.

Donostia San Sebastian Turismoa. (2020) https//sansebastianturismoa.eus.

Elias N. (1979 [1939]) *The Civilizing Process*. Oxford: Blackwell.

Fernandes E. (2009) *País Basco. A origem da modernidade da cozinha espanhola*. Lisboa: Ediçoes Do Gusto.

García-Hernández M., de la Calle-Vaquero M. and Yubero C. (2017) 'Cultural Heritage and Urban Tourism: Historic City Centres Under Pressure', *Sustainability* 9: 1–19.

González M. and Corcuera M. (2008) *La cocina donostiarra. Historia gastronómica de San Sebastián*. Donostia: Ediciones Al Gusto.

Hall M. and Sharples L. (2003) 'The Consumption of Experiences or the Experience of Consumption? An Introduction to the Tourism of Taste', in Hall M. and Sharples L. (eds) *Food Tourism Around the World: Development, Management and Markets*. Oxford: Butterworth-Heinemann: 1–24.

Hobsbawm E. J. and Ranger T. (eds) (1983) *The Invention of Tradition*. Cambridge: Cambridge University Press.

Lee D. (2007) '36 Hours in San Sebastián', *The New York Times*, www.nytimes.com/2007/11/18/travel/18hours.html?searchResultPosition=4.

Leizaola A. (2006) 'Matching National Stereotypes? Eating and Drinking in the Basque Borderland', *Anthropological Notebooks* 12 (1): 79–94.

Mak A., Lumbers H. N. and Eves A. (2012) 'Globalisation and Food Consumption in Tourism', *Annals of Tourism Research* 39 (1): 171–196.

Medina F. X., Leal M. del P. and Vázquez-Medina J. A. (2018) 'Tourism and Gastronomy. An Introduction', *Anthropology of Food* 13: 1–4.

Satran J. (2014) 'The 20 Most Michelin-Starred Cities in the World (PHOTOS)', www.huffpost.com/entry/best-restaurant-cities_n_928196?guce_referrer=aHR0cHM6Ly93d3cuc2Fuc2ViYXN0aWFudHVyaXNtb2EuZXVzLw&guce_referrer_sig=AQAAAMAluutwwx9U4NBxHR0XpVKqqbCRdJSY-OCrA1VndyfmW1-NNA3kc4BVhP_qfjO9CWu86lGEEbmUCADP6mqN0U6O7wZXjsJltguwWHlUj5igzFJVlYrZCQqZtGnGKBenp7sH7PNZPq-aMsrE8la5Uo4tmYaiprU5oHQEaYWHyy-i&_guc_consent_skip=1596730002.

Sutton D. E. (2010) 'Food and the Senses', *Annual Review of Anthropology* 39: 209–223.

Tourist Maker (n.d.) '10 Best Food Countries in the World', www.touristmaker.com/blog/top-10-leading-culinary-destinations/.

TW50B. (2020) 'The World's Best 50 Restaurants', www.theworlds50best.com/list/1-50.

Wolf E. (2020) 'What Is Food Tourism?' https://worldfoodtravel.org/what-is-food-tourism-definition-food-tourism/.

14 Source and supply

Situating food and cultural capital in rural–urban interactions in Vietnam

Catherine Earl

A sunny city laneway, car-body wide, paved hot. Motorbikes, silent, seats baking, a long row. Blocked houses, metal grill faces, windows open eyes. Children inside, outside their voices. Sunday. Extended families meet.

The gate, calling out, opens. My friend's front room, shoes off, tiles cool, floor seating, crowded, legs crossed, collars damp, heads turn, greetings. Brothers-in-law, yo! Toasting her husband, red faces, red eyes. Nearby, her mother-in-law, babysitting, crossed legs. The children, in her hair their fingers, on her thighs their feet, dancing. Behind them, my friend, smiling. The kitchen, steamy, sizzling, her sisters-in-law. Red faces, the electric fan not spinning, the food not cooling. Wet hands, the greens, rinsing, sorting, chopping, tossing, laughing. Her kitchen feeds her in-laws. For her, a test, a trial.

A sack struggling, startled, a shriek, surprised, a giggle. The sack, now walking, heading for the door. Intrigued, a warning, too late, a scolding. The kitchen a flutter. Each brace the orchard caught, not bought. Squawking feathery flapping, messing, resacked. Tiles wet, detergent slippery. The in-laws' meal, a regular. This kitchen, a first.

Hands over grandchildren's eyes, a song in their ears. Their mother, the birds, one by one cracked, hot water, head first, peeled, opened, emptied, their beaks, the sack, silent, still. Tiles wet, detergent slippery. My friend's initiation. Experiencing family, sharing.

Men toasted, pigeons roasted, yo! Smokey, sweet, salty, spice. Heads and feet, crispy, crunchy, moist, moreish. Bones in a pile, bottles in a row, children sleepy, time to go.

This scene, while unique, is typical of the banal commensality families across Ho Chi Minh City (HCMC) enjoy on Sundays. This chapter focuses on a neolocal family of four – a father, a mother, and two daughters in elementary school. The parents, both rural-urban migrants, own a house in a northern district of HCMC. The area developed from rice paddy fields in the 1990s to densely populated residential suburbs in the 2000s. This period of rapid growth was shaped by extended processes of macro-economic policy reform in the post-war years from the late 1970s (Fforde 2018). Rapid settlement of peri-urban agricultural land illustrates, firstly, the incredible scale of recent rural-urban migration flows to HCMC and, secondly, the growth of intra-urban mobility from the unaffordable inner city to less expensive peri-urban areas (Gubry *et al.* 2010). Migration experiences mark HCMC households. My friend's family is typical of families in HCMC; she and her husband are first-generation rural-urban migrants and, while their children are city-born, the family maintains close ongoing contact with their relatives in rural areas. Weeknights, they livestream meals via Zalo, Viber, or Facebook and eat together on weekends (Sundays) in each other's houses.

Like many of HCMC's small neolocal households comprising parents and one or two children, my friend's family hosts close relatives and guests on weekends. Some travel short distances from other city neighbourhoods; others make a longer journey from rural areas in the vicinity of HCMC. Rural visitors, such as my friend's brothers-in-law, may supply fresh or live produce, such as braces of pigeon, that they have grown, raised, caught, or sourced from trusted local suppliers outside the city. Food circulates with family members in a singular field of interaction that is neither rural nor urban but comprises an intermixing of practices and sense-worlds that blur boundaries between the rural and the urban. In this chapter, I deal with these as 'sensory interfaces', where different forms of sensory knowledge meet (Low and Kalekin-Fishman 2010).

Circulations of migrants and their meals across rural and urban contexts have a broader influence on urban food cultures. For rural-urban migrants, food in the city may take on new meanings. For example, culinary experiences could transform when ingredients that were plentiful in one location become rare or unattainable in the city or when dishes that were unknown outside the city become familiar, perhaps favoured, in a more diverse culinary landscape. Particular culinary experiences may be transported between locations and shared with relatives and others in different places. There, they may be further transformed and appropriated in other food cultures. Such processes speak to heterogeneity of food cultures in contemporary Vietnam and remind us it is potentially misleading to assume a link between place, culture, and food culture (Avieli 2005: 168).

A dish does not belong to a place. Rather, it belongs to the senses; it is situated in its practice. A dish moves to new places and in its mobility, it potentially transforms: each pigeon meal comprises different pigeons, good season birds for example, a different amount of charcoal, and the cook's 'signature' spice blend. Such differences shape the experience of the senses. A dish is part of a meal shared by fellow diners, a family on Sundays. Commensality and memory shape belonging in the family and re-inscribe kin ties. Through shared food preparation and consumption, the senses of the cooks and diners may be cultivated.

Cultural studies have long established that, wherever they originate, familiar and authentic practices of food sharing can be a significant means of cultural expression (Warde 1997: 22–23). These can be recognised by others as indicators or markings of embodied cultural capital that convey enduring symbolic value in a particular socio-historically situated context: for example, a Sunday family meal. In this chapter, drawing on culinary sensory interfaces of HCMC,

I argue that the phenomena of embodied cultural capital that distinguish urban middle classes from others are not primarily acquired from or located in the city. Rather, since embodied cultural capital takes the form of 'indelible marks' on the bodies of those who possess it (Bourdieu 1984: 79), like them, cultural capital is mobile and can be sourced and accrued in diverse places. Moving beyond a focus on the metropolitan (for example, Bélanger *et al.* 2012), I argue the stuff of distinction circulates in a singular field of interaction that traverses rural and urban locations. This singular field of interaction shapes and is shaped by both rural and urban sensory experiences and is characterised by an intermixing of heterogeneous sense-worlds. In *Vietnam's New Middle Classes*, I drew on Bennett *et al.* (2009), among others, to point out that a new aesthetic language of urban leisure lifestyling among middle classes of HCMC is characterised by fluidity, heterogeneity and cultural dissonance that, for example, values crossover tastes which merge a changing selection of rural, globalised, high-brow, and low-brow influences (Earl 2014). In the following discussion, I extend this argument to emphasise that, firstly, the singular field of interaction in which a new aesthetic language of middle-class urban lifestyling circulates is relational not dichotomous, and thus, it is unconscious of a rural-urban divide. Secondly, I illustrate that embodied cultural capital associated with culinary-oriented sense-worlds is characterised by an 'indifferentiation' (Lahire 2008) that smooths over assumed dichotomies of low-brow and high-brow, rural and urban, traditional and modern tastes. In this approach, like Mark Liechty (2003), I locate middle-class experiences in the 'middle'.

Methodologically, this chapter draws on multi-sited ethnographic fieldwork carried out in Vietnam since 2000. Inspired by the interdisciplinary mobilities turn of the social sciences (Sheller and Urry 2006), I deploy mobile methods in my research among mobile subjects in and of HCMC. My approach draws on the first and second of twelve mobile methods outlined by Monika Büscher *et al.* (2011: 8–9) that privilege participant observation with people in their patterns of movement. I experience co-present immersion as I 'follow the people' (Marcus 1995: 105) across HCMC and beyond its physical boundaries (see Earl 2008, 2014). Moreover, I expand participant observation that privileges ocularcentrism to 'participant sensation', which incorporates not just sight but all the senses.

In a project centred on phenomenal and sensory experiences, my methods move beyond a normative discursive paradigm centred on the collection of personal narratives. That is, I explore what people do, instead of recording what they say they do. For example, I share the experience of preparing and consuming a pigeon meal; I don't record my friend saying in person or posting on social media that she is preparing or has prepared a pigeon meal. Dara Culhane (2017: 52) describes sensory ethnography, a form of anthropology of the senses, as moving away from the '"verbocentric" and "textcentric" academic traditions that mitigate against recognising forms of sensory knowledge that resist representations by written word'. In this chapter, I join the family of four to explore the contrasting sense-worlds and commensality they experience across their three homes: firstly, the neolocal home in the north of HCMC where the family live; secondly, their patrilineal homeland (*quê nội*), the place where the father was raised; and thirdly, their matrilineal homeland (*quê ngoại*), the place where the mother was raised. Using words to capture sensations, I share with readers how it felt, not what was said.

Sensory experience, commensality, and cultural capital

The opening anecdote highlights the cultivation of sensory experience in migrant food cultures of HCMC in a sensory interface, a space where the senses meet. Sharing a meal of pigeon and other examples of commensality show that culinary-oriented sense-worlds, while shared, are individualised; individuals experience their own sensations. The senses collaborate, Maurice Merleau-Ponty (1962) asserts. Thus, as David Howes (1991) emphasises, while the senses can also conflict, a correlation of sensations gives a meal its flavour. In participant sensation, mouth feel and flavour are not more important than sight, smell, and sound. A Vietnamese meal, such as spice roasted pigeon, is delicious (*món ngon*) when the five senses correlate, and the individual senses are multiplied: taking in a balance of pleasing aromas (smell); contrasting colours (sight); contrasting soft, smooth, chewy, moist, fresh-crunchy, and crispy-crunchy textures (mouth feel/touch); crunchiness (sound); and complementary sweet, salty, sour, bitter, and hot/spicy elements (flavour/taste). Learning how to prepare a Vietnamese meal involves learning how to provide the experience of the correlation of sensations to each diner: for instance, as a new wife preparing a meal of the patrilineal family's familiar cuisine. In a family home, Vietnamese rarely ask, 'How are you?' or 'How was your day?' Rather, they ask, 'Have you eaten?' Producing, supplying, preparing, and sharing meals among Vietnamese is a way of demonstrating togetherness and expressing bonds of parental/fraternal love (*thương*) in family life.

Showing love in a family involves a process of cultivating the senses. The pigeon meal reveals my friend's commitment to consciously cultivating her senses of the kitchen and table. Although she did not grow up eating pigeon, she demonstrates she has learned how to enjoy a meal of pigeon with her husband's family. Through demonstrating her ability to prepare and serve an authentic pigeon meal to her in-laws – long-time pigeon eaters – she embodies new dispositions. She has learned how to recognise specific high-quality ingredients familiar to them and how to prepare an authentic meal for them to share. Preparing and serving enjoyable meals is how she shows them love. Her in-laws' willingness to eat her food is how they show her love. New, learned culinary-oriented sensory experiences and skills do not displace her existing awareness of food quality, cooking ability, or tasting palate. Rather, these expand what she can do and what she knows in her bodily practices. These practices can be recognised by others – her in-laws, their kin, and their social circle – as an achievement of the highly valued signals of cultural competence or, in the language of Pierre Bourdieu, embodied cultural capital.

Cultural capital is a central concept deployed by Bourdieu in his detailed exploration of class-divided societies. In its embodied state, cultural capital takes the distinctive form of long-lasting dispositions evident via inscription on the body: for example, in regional pronunciation or culinary palate (Bourdieu 1997: 48–49). Among the culturally inscribed embodied practices, or 'indelible' marks of childhood (Bourdieu 1984: 79), are the familiar sensations of both mundane and special family meal preparation and consumption that my friend acquired through direct contact and ongoing exposure. This illustrates that the transmission of dispositions and moral ethics do not simply 'trickle down' from one generation to the next so that societies become 'shackled' to traditions (Daloz 2008: 315). Even when inherited, embodied cultural capital and other dispositions must be cultivated.

Cultivating the body's abilities is conscious: 'To change a body of habits, physical or cultural, can never be a matter of wishful thinking and trying; it depends on learning and practicing new techniques' (Jackson 1989: 119). Thus, the efficacy of embodied cultural capital is not an effect of natural aptitudes. Rather, it is shaped by a willingness to invest time and other resources, including other forms of capital, for an individual to consciously develop embodied practices that become inherent, or 'indelible', and which are recognisable to others as marking cultural competence and even social distinction. For example, learning how to handle and prepare a brace of pigeons when one is familiar with preparing a single chicken.

In a Vietnamese social world, Philip Taylor (2016: 1) argues, practice-oriented social relationships are not governed by a universal rationality or unalterable cultural template. Cultural homophily cannot be taken for granted. For Taylor, there are tangible and intangible relations of hierarchy and reciprocity. There are modes and intensities of connections and disconnections that vary according to class, region, gender, and ethnicity. These and other social structures and boundaries are mediated through sensory experiences that occur at interfaces where different forms of sensory knowledge meet. Through social interaction, sensory experiences are imbued with social meaning and symbolic value. Sharing a Sunday lunch of wild-caught pigeon with visiting in-laws is a case in point. At this place, relatives and guests have opportunities not only to experience the senses collaborating or conflicting but also to express the cultivation of their senses as cultural competences or skills that are recognised in that time and place as marking relative social position.

Bourdieu (1997: 49) points out that various profits of forms of capital are realised because economic and cultural resources are not distributed evenly, and agents do not share equal means to exercise social power. For example, access to economic capital enables the privatisation of the sense-world and the deployment of social power to choose which forms of sensory experience to multiply and which to dull or avoid (Earl 2018). Thus, the indelible marks of distinction are not fixed. 'Rubbing shoulders' with others, Bernard Lahire (2008: 174) contends, transfers heterogeneous marks or residues of varying strengths and sometimes contradictory dispositions. Moreover, the social meaning and symbolic value attached to marks of distinction are not universal; they are place-dependent within that context and socio-historically situated.

As I argued previously, there is a fluidity, heterogeneity, and dissonance of cultural tastes that characterise a new aesthetic language of urban leisure lifestyling among middle classes in HCMC (Earl 2014). Migrant experiences at culinary-oriented sensory interfaces most evocatively capture the myriad ways of experiencing and valuing HCMC sense-worlds and illustrate that cultural homophily cannot be taken for granted.

Experiencing the migrant sense-worlds of HCMC

While many migrants arrive to and exit from HCMC in seasonal flows, others settle more-or-less permanently in the developing mega-urban region. A desire for a more comfortable and secure life in the future for oneself and one's family is a key motivator in migrant decision-making. The paternal grandparents initially fled war in central Vietnam in the 1960s to resettle in the rural south-central coast where the father was born. In part, their initial migration as refugees was made possible through kinship networks as well as other social connections and relationships. In the new place, they joined their previous neighbours. Having established themselves as orchardists, they sent all their children to school, one child in each class until the older ones were working and could support the younger ones to enter high school. The father, one of the younger siblings, was able to attend university in HCMC. After he graduated, he remained in the city for work, married, and settled there.

The mother's story is a similar one, although that of her family differs. The family were settlers, having arrived in the Mekong Delta generations earlier. While their origins are distant, the ancestral graves the family tend are nearby. The children were educated at the village school, year after year, until the younger ones were supported by the older ones to make the move to HCMC for university study. After the couple married, like many young professional couples, they bought a house in the expanding peri-urban fringe of HCMC to raise their children – two daughters – at the time of fieldwork in first and second grade. In their city home and the rural homes of their parents, the family experiences contrasting sense-worlds.

Neolocal residence in HCMC

Inside the house are the daughters' games, their laughter, bickering, crying, their mother's scolding, their chanted school lessons, their words merging to a buzz as they race to finish the page. The wet season, a downpour, the lightwell window leaking, drips become plops, the sizzle of hot oil in a wok, the click of the rice cooker finished, the crack of eggs for their supper, the crunch of crispy snacks (if their mother is not looking). The washing machine clanks, the television laughs, the air-con whirs on, off, their father's voice: 27 degrees – maximum!

There is cooking gas, chilli on the air, fresh steamed rice, fish sauce, raw garlic. Soap, the girls' washing. Insecticide. Motorbike fumes: their father returns, their mother leaves for class. Neighbours' meals, neighbours' drains, and sometimes oversweet fruit on the ancestral altar.

Tastes in the city are fleeting, the girls fussy. Simple stir-fries, meat-fish-greens for one, omelette for the other. The meals quick, then green guava with chilli salt, their mother's favourite. If not, the village may have sent a durian, a jackfruit. Hot tea for adults.

The house is either hot or cool. Outside bright, hot concrete. Inside dark, cool tiles. Water on the tiles, slippery, soapy. Drips from the ceiling, splashy, dirty. Don't touch! Hot stove in the kitchen, cool water in the bathroom. Rain outside, hanging clothes brush you past. Dry laundry, wet laundry, dirty laundry. Downstairs stuffy rooms, a fan for meal times, cooling the rice, the tea. Upstairs unpolished floorboards, at the front maybe a breeze, at the back the girls' room air-conditioned: 27 degrees.

Father's village: **quê nội**

The father, now a metropolitan middle manager, was born into a large farming family in a south-central coastal village, where daily life centres on the orchard and the beach.

Near the house, chickens, ducks, dogs, rats, and the wind whistling. A machine generating, a large motor transporting a crop or farm supplies, a small motor transporting an uncle-worker, a truck reversing, another taxi arriving from the train station. Guests in the house. The metal frames of glass windows opening and closing at the front, the slap of wooden shutters at the back. Water into a ceramic urn, low-pressure tinkle and drip, drip, drip. A pump starting. Water running and echoing. An urn filling and spilling. Food parcelled in leaves brittle, leaves fresh. Plastic bags soft, canisters hard. A guitar, singing, a bicycle bell, ringing, the phone, the guttural vowels of the local accent.

There is fertiliser and late season rot, salt and seaweed, incense for the ancestors, candle smoke, an outdoor cooking fire, wood smoke, burning refuse, chemical smoke, diesel for the generator, toffee for puffed rice candy.

The taste of the orchard is hot steamed rice, sour pickles, fish sauce, whisky, strong coffee, minty toothpaste, no more candy.

Gritty, a stone seat, eating outdoors, on the table leaves falling and ants crawling. Smoke in the eyes. A sister's pinching. The bicycle, one on the seat, the other on the carrier rack, four feet on the pedals, wind in the hair, a scraped knee, a sticky plaster. To the beach, swimming, going under, water up your nose, sand in your clothes, seaweed slimy on your toes. A shell to your ear, a crab in your hand. Hot asphalt underfoot. Sunburn, cooling gel, an ice cream treat.

Mother's village: **quê ngoại**

The mother, an office administrator in HCMC's industrialised fringe, was born into a large farming family in a Mekong Delta village, where daily life revolves around rice farming, lotus cultivation, and lay Buddhism.

There are palm leaves rustling, canal fish plopping, a motorbike passing, the grandfather's bicycle wheel grating, 'hmph' when he is waiting. The kettle whistling, coconuts thudding, insects buzzing, chickens clucking, the rooster not waiting for dawn, the clock reliably chiming. A tap filling a plastic bucket, bath water sloshing on concrete, the girls splashing and laughing and complaining. A dog barking, the landline phone ringing, the loose vowels of the local accent.

A celebration starting, a motorbike arriving. Another, many more. Karaoke, an emcee, buzzing speakers. Clicking heels, a singer, a duet, a diva. An uncle drunk, crooning, slurring, a toast – yo! Another, many more. Children's card games, laughing. One crying, boys running. Kicking, cheering, a football skimming. A splash, gasps, sighs, their disappointment. Bowls empty, metal tapping, melamine clacking, chopsticks dropping. Water sloshing, women joking, guests leaving. A broom on concrete, aluminium cans, glass bottles, bottle tops. The babies waking, feeding, the household sleeping.

Early morning incense paying respect to the ancestors. Flowering lotus, ripening jackfruit. Canal water, rice paddy water, washing water, laundry soap. Burning leaves, burning plastic, sometimes gas on the air. Body odour, steamed rice, chilli, lime, cigarette smoke, beer, whisky. Mouths of sweet, sour, bitter, salty, and spicy.

The textures and their contrasts are intense. Rice too hot, ice too cold. House too dark, its air too still. Sun too harsh, paving too hot. Football too hard, cousins too fast. Feeding fish. The dyke too dry, too rough. Stumbling. The bank too wet and sucking. Collecting coconuts. The grass too high, itchy arms and eyes, a sneeze. Harvesting vegetables. Insects too many. Crawling, flying. A scream. Another. Both crying, a cuddle, clinging on. The younger refusing to set her feet down, heavy, her fingernails sharp. Me afraid she will slip from my hip. Will she be hurt? Will she (or we) slip from the bank into the sticky mud, the mucky pond? Then, solid path, paved terrace. Resting in the hammock. Red king ants walking, biting. Her mother's kisses no relief.

Commensality and situating migrant food cultures of HCMC

The experiences of the city-born girls across their three homes illustrate that cultural capital of culinary-oriented sense-worlds is mobile. City residents may be exposed to rural experiences without being in or of a rural area. Regular family trips across and beyond the city are typical activities of HCMC. These include undertaking mundane leisure outings as well as special trips to one's own or someone else's ancestral homeland (*về quê*), such as for life-course milestone feasts such as death anniversaries (*đám giỗ*), wedding feasts (*hôn lễ*), or first birthdays (*đầy tháng*). Migrants returning to their ancestral homes typically invite their friends, colleagues, and classmates to join in, to accompany them on the journey, and to share the meals served at the destination. Other times rural visitors come to the city. Mundane and special visits alike offer sensory interfaces where different forms of sensory knowledge meet. At sensory interfaces culinary-oriented sense-worlds may be disrupted, broadened, multiplied, or diversified by exposures to new or unfamiliar sensory experiences and tastes. Some of these exposures may be pleasant and relaxing, such as listening to palm leaves rustle or enjoying a home-grown fruit, and others may be uncomfortable, such as hearing excessively loud music or being bitten by insects. Certain sensations, such as listening to a rooster crow, may be pleasing to some and annoying for others.

These examples I show are not located in either a rural or an urban context. Sensory interfaces are mobile; they move with the sensing bodies of the participants. When participants quit a place, the sensory interface they share also goes. The opening anecdote showed a pigeon meal that is often shared by the family in the natal home of the parents (and also the urban home of other siblings) being shared at the couple's city home. The previous section conveys the contrasting sensory interfaces of the parents' village homes to illustrate the heterogeneity of Vietnamese sense-worlds and the diversity of sensory experiences with which the family is familiar and which, for them, while diverse and contrasting, are mundane. Even so, each family member has their own sensory experiences. What one parent experiences as familiar from childhood, the spouse and children have acquired through direct contact and ongoing exposure as they cultivate their senses over time and with effort. Furthermore, at the three homes, some sensations are dulled, others multiplied. Their bodily circulations, with the transfer of their emotions, dispositions, cultural goods, and commodities, occur within a singular field of interaction that is not conscious of rural-urban boundaries (Earl 2014).

Heterogeneity and diversity of experience characterise Vietnamese sense-worlds. A blurring of an assumed dichotomy of urban and rural has been noted in the geography and anthropology of Vietnam. Jamie Gillen (2016), for example, argues that everyday practices and imaginative discourses of rurality in HCMC are not displaced in urbanisation processes, but rather, city residents incorporate rurality into urban experiences. With frequent circulations between rural and urban locations, intrusions of rurality and urbanism are not only of the city. Ashley Carruthers and Dang Dinh Trung (2018), in their study of return migration from HCMC to a rural village, demonstrate that social, emotional, and material resources flow both ways between the city and the village in exchanges that are horizontal, not hierarchical. A return to the village, a place of greater social mutuality and care but lower socio-economic standards and fewer career opportunities, may not exemplify downward mobility and failure. Exposure to differentiated sense-worlds, whether located in rural or urban places, may transform embodied practices in ways that are recognised by others as valued forms of capital. As the culinary-oriented sense-worlds of HCMC in this chapter show, there are limitations in conceptualising village living on the one hand and city living on the other. I locate sense-worlds in bodies that cultivate sensory experiences. When the people in these anecdotes move between locations and across sense-worlds, they encounter a diversity of culinary-oriented sensations that share a growing 'indifference' to low-brow and high-brow, rural and urban, traditional and modern orientations, particularly when they are closely associated with authenticity and quality. The following examples of commensality at the family's three homes illustrate the indifference of highly valued cultural signals to dichotomised constructions of place.

Bamboo and a bonfire: **quê nội**

The father's village, night. A feast next door, we walk. The road, mud now set rutted and rough. Stumbling. The house wooden, shutters open. New guests, more stools, bowls, chopsticks. Squash in, music. Our corner, dark. Wire, pliers, a fluoro tube. More light, bright. Beer or cola? My bowl, pickles. Brown, firm not soft, smooth not crunchy, jerky not meat, bland a plant. Bamboo. Home-grown, homemade. Whose home? Heads turn, she waves. How made? Large crocks, each year. To her, a toast! Bowls empty, tables clear. Sky black and stars. Sea breeze cold, bonfire hot. Guitars strum. Wood smoke or mosquitoes? Songs sing folk, words all voices, shoulder to shoulder, swaying, the next tune.

Lotus stem and shrimp: **quê ngoại**

The mother's village, a celebration. Brothers and sisters, wives and husbands, children, neighbours, colleagues, classmates. So many faces, too many names. Blistering sun. Karaoke. Concrete terrace, marquee shade. Big round tables, lightweight stools. The starter, a salad. Fragrant and fresh, lotus stem. Across the path, the pond. Knee-deep water, mud feet. One small knife, bend. White, a straw crisp hollow. Slice lengthwise, rinse out the mud. Carrot, coriander leaf, peanuts, sour dressing. Table clear, reset. Hexamine stove, chemical nose, a burnt sleeve. Wide low pot, the lid lifting, dropping. A clatter. The heat, a frenzy, a claw escaping, still blue. Chattering, clacking. Hotter still and silence. Now open, scented steam. Red and crisp, black eyes, still and stare. Yellow brains, sweet meat. Shells crunch under feet, beer on our breath, cigarettes in our eyes, dry throat, songs in our ears.

Pigeons: neolocal home

The family's laneway, Sunday, sunny day. A sack brings motorbikes. The kitchen alive. Grandmother, knees and hands, singing to the children. Uncles toasting, aunts roasting, sweaty faces. Feathers and bones. Father, mother, together, family. Small birds, long lunch.

Conclusion

In this chapter, I argued that the phenomena of embodied cultural capital that distinguish urban middle classes from others are not primarily acquired from or located in the city. Rather, since embodied cultural capital takes the form of 'indelible marks' on the bodies of those who possess it (Bourdieu 1984: 79), like them it is mobile and can be sourced, accrued, displayed, and recognised by others in or of a range of locations. Vietnamese culinary-oriented sense-worlds illustrate that there is not a universal set of cultural cues that indicate middle-classness or other enduring social status. Highly valued signals that may be recognised by others as cultural capital demonstrate a cultivation of the senses, which may include but is not reducible to embodied practices. Importantly, the cultivation of the senses is dynamic, not fixed, and it is relational since it is shaped by the recognition of others.

In the earlier sections, I explored contrasting sensory interfaces experienced by migrants to illustrate that, firstly, the singular field of interaction in which a new aesthetic language of middle-class urban lifestyling circulates is unconscious of a rural-urban divide. I argued the stuff of distinction circulates in a singular field of interaction that traverses rural and urban locations and blurs cues that point to distinct rural or urban phenomena.

Secondly, my analysis of sensory interfaces illustrated that embodied cultural capital associated with culinary-oriented sense-worlds is characterised by heterogeneity, diversity, dynamism, and an indifference to a set of fixed cultural cues. The daughters, like their father, are second-generation migrants who were born in a new place away from their ancestral homeland. But, while the children are from and of the city, as they are raised and grow up, they regularly experience the diverse sights, sounds, smells, tastes, and touch-textures of their parents' village homes. The contrasting sensory interfaces expose the children to the mundane sensations of three qualitatively different homes. Their parents' experiences are contrasting since they faced consciously learning how to live in the city as well as in each other's villages. The mother's new experiences in the patrilineal family enabled her to consciously cultivate her senses in doing more than learning how to eat new dishes; she cultivated familiarity with the broader contexts of food production, including safe sources and secure supply, that conveyed symbolic value to enhance her underlying reservoir of embodied cultural capital.

The profits of migration are not located in a single location, and they do not flow in only one direction. While metropolitan contexts provide wider opportunities to experience cultural phenomena, rural people may be at a relative advantage over urban people in accruing a reservoir of enduring embodied cultural capital derived from a diversity of authentic cultural resources, so long as these retain value in the particular (not universal) place-dependent and socio-historically situated contexts in which their cultural achievements can be recognised by others as forms of distinction.

References

Avieli N. (2005) 'Vietnamese New Year Rice Cakes: Iconic Festive Dishes and Contested National Identity', *Ethnology* 44: 167–187.

Bélanger D., Drummond L. B. W. and Nguyen-Marshall V. (2012) 'Introduction: Who Are the Urban Middle Class in Vietnam?' In Nguyen-Marshall V., Drummond L. B. W. and Bélanger D. (eds) *The Reinvention of Distinction: Modernity and the Middle Class in Urban Vietnam*. Berlin: Springer: 1–17.

Bennett T., Savage M., Silva E., Warde A., Gayo-Cal M. and Wright D. (2009) *Culture, Class, Distinction*. London: Routledge.

Bourdieu P. (1984) *Distinction: A Social Critique of the Judgement of Taste*. London: Routledge and Kegan Paul.

Bourdieu P. (1997) 'The Forms of Capital', in Halsey A. H., Lauder H., Brown P. and Stuart Wells A. (eds) *Education: Culture, Economy, and Society*. Oxford: Oxford University Press: 46–58.

Büscher M., Urry J. and Witchger K. (2011) 'Introduction: Mobile Methods', in Büscher M., Urry J. and Witchger K. (eds) *Mobile Methods*. London: Routledge: 1–18.

Carruthers A. and Dang Dinh Trung. (2018) 'On the Myth of Uncivilised Rural People in the City', in Earl C. (ed) *Mythbusting Vietnam: Facts, Fictions, Fantasies*. Copenhagen: NIAS Press: 163–181.

Culhane D. (2017) 'Sensing', in Elliott D. and Culhane D. (eds) *A Different Kind of Ethnography: Imaginative Practices and Creative Methodologies*. Toronto: University of Toronto Press: 45–67.

Daloz J.-P. (2008) 'Towards the Cultural Contextualization of Social Distinction', *Journal of Cultural Economy* 1: 305–320.

Earl C. (2008) 'Longing and Belonging: An Ethnographic Study of Migration, Cultural Capital and Social Change Among Ho Chi Minh City's Re-Emerging Middle Classes', Unpublished PhD Thesis, Victoria University, Footscray, Australia.

Earl C. (2014) *Vietnam's New Middle Classes: Gender, Career, City*. Copenhagen: NIAS Press.

Earl C. (2018) 'Senses of Distinction: Social Differentiation, Mobility and Daily Life in Ho Chi Minh City', in Low K. and Kalekin-Fishman D. (eds) *Senses in Cities: Experiences of Urban Settings*. London: Routledge: 39–54.

Fforde A. (2018) 'Myths in the Understanding of the Process of *Đổi Mới*', in Earl C. (ed) *Mythbusting Vietnam: Facts, Fictions, Fantasies*. Copenhagen: NIAS Press: 189–210.

Gillen J. (2016) 'Bringing the Countryside to the City: Practices and Imaginations of the Rural in Ho Chi Minh City, Vietnam', *Urban Studies* 53: 324–337.

Gubry P., Le T. H., Tran T. T. T., Nguyen T. T., Pham T. H. and Vu H. N. (2010) 'Intra-urban Mobility in Ho Chi Minh City and Hanoi', in Gubry P., Castiglioni F., Cusset J.-M., Nguyen T. T. and Pham T. H. (eds) *The Vietnamese City in Transition*. Singapore: ISEAS: 63–100.

Howes D. (1991) 'Sensorial Anthropology', in Howes D. (ed) *The Varieties of Sensory Experience: A Sourcebook in the Anthropology of the Senses*. Toronto: University of Toronto Press: 167–191.

Jackson M. (1989) *Paths Toward a Clearing: Radical Empiricism and Ethnographic Inquiry*. Bloomington: Indiana University Press.

Lahire B. (2008) 'The Individual and the Mixing of Genres: Cultural Dissonance and Self-Distinction', *Poetics* 36: 166–188.

Liechty M. (2003) *Suitably Modern: Making Middle-Class Culture in a New Consumer Society*. Princeton, NJ: Princeton University Press.

Low K. E. Y. and Kalekin-Fishman D. (2010) 'Afterword: Towards Transnational Sensescapes', in Kalekin-Fishman D. and Low K. E. Y. (eds) *Everyday Life in Asia: Social Perspectives on the Senses*. Farnham: Ashgate: 195–203.

Marcus G. (1995) 'Ethnography in/of the World System: The Emergence of Multi-Sited Ethnography', *Annual Review of Anthropology* 24: 95–11.

Merleau-Ponty M. (1962) *The Phenomenology of Perception*. London: Routledge and Kegan Paul.

Sheller M. and Urry J. (2006) 'The New Mobilities Paradigm', *Environment and Planning A* 38: 207–226.

Taylor P. (2016) 'Introduction: An Overture to New Ethnographic Research on Connection and Disconnection in Vietnam', in Taylor P. (ed) *Connected and Disconnected in Viet Nam: Remaking Social Relations in a Post-Socialist Nation*. Acton, ACT: ANU Press: 1–40.

Warde A. (1997) *Consumption, Food and Taste: Culinary Antinomies and Commodity Culture*. London: Sage.

15 Preparing *Uchu Jaku*

The politics of care in a traditional Andean recipe

Paz Saavedra, J. Guillermo Gómez-Urrego, and José David Gómez-Urrego

It is a wet and cold Wednesday morning in Paquiestancia, a small rural town located in the Andean canton of Cayambe in Ecuador. *Doña* Luisa, an 80-year-old woman, and her granddaughter Alicia (pseudonyms provided), a woman in her 30s, are packing *Colada de Uchu Jaku* in large metallic milk containers to keep it warm. *Colada de Uchu Jaku* is a creamy flour-based soup typical of the area, and the artefact through which we conducted our ethnographic study. *Doña* Luisa and Alicia will transport the hot soup five kilometres from their rural hometown to the city of Cayambe to sell in the *feria* Bio-Vida, an agro-ecological market. On this occasion, the two women were not able to chargrill the *cuyes* (guinea pigs) for the soup as they would normally have done under less damp weather conditions. Instead, they fried them inside the kitchen protected from the persistent rain and wind.

Cayambe, both its urban and rural areas, is the ancestral territory of the indigenous nationality Kichwa Kayambi. *Uchu Jaku* is a Kichwa name that can be translated as 'spicy flour' because the flour for the soup originally contained chillies along with seven roasted grains and other condiments. Although flour-based soups are common across the Andes, this one is unique for its numerous ingredients as people originally cooked it for celebrations and festivities. It contains lamb's meat for the broth that pairs nicely with aromatic cumin and garlic. The soup is then topped with a refreshing slice of fresh cheese, a succulent quarter of juicy roasted *cuy*, and half a boiled egg. The *cuy* corroborates the soups' celebratory character as *cuy* is not eaten every day in this territory. The dish is accompanied by homemade *ají* (hot sauce) that warms up diners immediately in the cold early Andean mornings.

Doña Luisa and Alicia form part of a group of agroecological producers from different rural localities across the Northern Andean region of Ecuador who gather every Wednesday at the *feria* to sell their fresh products. The Bio-Vida *feria* sells a wide variety of fresh organic products and cooked dishes. The producers also take turns preparing hot food for sale. The soup is part of the *feria*'s weekly menu, and this week was *Doña* Luisa's group's turn to cook. The selling of hot food in the *feria* started with the producers collecting traditional family recipes. *Colada de Uchu Jaku* was one recipe recalled by the families, and

Doña Luisa is one of few producers who have a family recipe for the *Colada*; she learned to prepare it with her mother.

This chapter discusses the labour of care and the politics of rural farmers who sell traditional food in the city. We follow Doña Luisa and Alicia across different spaces and sensorial experiences of preparing, transporting, and selling *Colada de Uchu Jaku*. Our goal is to discuss and make visible the labour of care involved in adapting the traditional food of rural Andean livelihoods to fit the rhythms and regulations of selling food in the city. Through the story of the *Colada de Uchu Jaku*, we discuss how agroecological farmers supply fresh organic food to the city while continuing the city's connection to ancestral traditions of the Kayambi territory. We have chosen *Colada de Uchu Jaku* for three main reasons: 1) it is a meaningful dish for the producers as they identify with the *Colada* as emblematic for the area; 2) the labour of care involved in its preparation showcases the work producers do to keep the *feria* running where this dish encapsulates diverse products, skills, and knowledge of local agroecological production; and 3) it shows how traditional dishes transform to keep them alive as the *Colada* has been adapted from being a traditional celebratory dish from the rural areas to a commercial dish in the urban *feria*.

The research draws on three main methods. The first is qualitative in-depth interviews: two interviews with *Doña* Luisa and two with her granddaughter. Second, participant observation: following them throughout the whole process of preparation and commercialisation of the dish. Third, constant visits to the *feria* over a period of six months. We also draw on Saavedra's PhD research on the *feria*, BioVida, and the producers working there, through interviews with key actors and observations conducted for seven months in 2017 and 2018.

The framework of care and temperature

The chapter draws on feminist studies of care. Following Joan Tronto (1993: 103), we define care as 'everything that we do to maintain, continue, and repair our "world" so that we can live in it as well as possible'. Under this definition, agroecology represents a holistic practice of care in which cooking represents caring in practice (Mol 2002). Through cooking, agroecological producers maintain ancestral knowledge by providing nutritious meals to their families while connecting consumers at the *feria* to the principles and values of agroecology.

Feminist political readings of care have the potential to make visible, disrupt, and re-imagine how care is performed and distributed (Puig de la Bellacasa 2017; Tronto 2010). We use these conceptual tools to make visible the work of the producers through the caring activity of cooking. Importantly, the framework of care allows us to examine complex, nonlinear, and conflicting practices (Duffy 2007; Roberts 1997) involved in preparing and selling the dish. Feminist scholars warn against a simplified association of care with a positive-value and loving activity. Broadly, they argue that care can reproduce oppressive dynamics that need to be challenged (Federici 2004, 2018; Murphy 2015; Raghuram 2016).

Furthermore, literature shows how care work falls unevenly on women, specifically women in the third world, working-class women, and women of colour (Hill Collins 2000, 2007; Roberts 1997; Salazar Parreñas 2015). Women from Latin American rural households are overwhelmingly in charge of cooking and taking care of their family crops (Minga 2014). In particular, women from rural Ecuador are reported to disproportionately use their time for caring activities in comparison to men, which represents a heavy workload for them (INEC 2012; CARE Ecuador 2016; Ortega 2012). In resonance with such tendencies, women at the *feria* prepare the *Colada* and are also the majority of the producers at the *feria*. Agroecology represents an opportunity for these women to take better care of their families by providing more diversified and nutritious diets while also generating an additional income.

In this chapter, we apply temperature as a lens to explore conflicting aspects of care present in the embodied experiences of the producers. By focusing on temperature, the chapter links food and the senses through the producers' experiences of heat and cold, rather than focusing on the more studied sensorial experience of eating. In this way, we echo research on the social and political realms of food that connect it to particular bodies, labour, and territories (Grey and Patel 2015; Whyte 2016; Grey and Newman 2018). Furthermore, we bring to the table the demanding and painful aspects of care in cooking that can be analysed through multiple sensorial experiences of temperature.

The Bio-Vida agroecological *feria* and *Doña* Luisa

Bio-Vida was established in 2007 with almost 70 founding members working collaboratively to support a model of agroecological production that uphold goals of fair trade, solidarity, and social justice. Agroecology supports small-scale family production using diverse crops that are free from chemicals to enhance families' diets and livelihoods. Additionally, it promotes farmer-to-farmer learning and collaboration with experts to meet local needs (De la Cruz 2018; Gortaire 2017; Siliprandi 2015). Each week, different groups take turns preparing the menu of typical dishes and drinks. *Doña* Luisa´s group is one of the few that prepares hot food, and she is in charge of the famous *Colada de Uchu Jaku*.

The day we visited *Doña* Luisa at the *feria*, it was pouring with rain. The *feria* takes place in the patio of a patrimonial building administered by the city council of Cayambe. Until recently, this was an occupied space where, only earlier this year, the council granted them formal access to the site. Indeed, the lack of formal support from the council denied the producers access to potable water and electricity. Moreover, when the rain is heavy and continuous, producers relocate their stalls around the corridors of the building protected by the roof. As the patio where the stalls are located is unpaved when it rains, everything gets muddy. After years of protest, several agroecological organisations in Cayambe, including Bio-Vida, succeeded in getting a law approved to regulate the use of public space for agroecological markets in the city. While its impact still waits to be seen, this achievement constituted a great feat in the struggle to ensure that public regulations include support for agroecological practices.

Agroecology works with local, ancestral, community, and technical scientific knowledge in a dialogue between knowledges and practices (De la Cruz 2018; Gortaire 2017; Siliprandi 2015). It seeks to transform production into a cycle in which every element serves a purpose, and the soil, water, seeds, animals, and humans are all adequately nourished. While agroecology has been an expanding movement in Ecuador since the 1980s, agroecological producers often work without appropriate infrastructure, legislation, and investment from local and national governments. Agroecological initiatives function against relentless land grabbing and contamination and accumulation of resources, such as contamination of water and fertile soil by agro-industry (Daza 2018; Larrea and Greene 2018; Macaroff 2018, 2019). Moreover, agroecological producers are part of the various social movements demanding an agrarian reform of property and land rights. However, agro-industry constitutes a strong political force in the country, influencing crucial public policy that is detrimental to small-scale agriculture (Sherwood and Paredes 2014). Without a reform that modifies this situation, pressure is placed on agroecological producers to overcome adverse conditions alone. This workload is particularly heavy for women because women's workload in agroecology relates not only to productive and domestic labour but also to political work and activism (CARE Ecuador 2016; CPES 2011).

Agroecological producers from Bio-Vida relate food to their fight for justice and sovereignty over their territory. We argue that this pursuit for a fairer model of production should not be analysed separately from the embodied and sensorial experiences of producers through which their work becomes meaningful. Producers find this work satisfying, proudly sustaining food memories across generations while appreciating how customers enjoy their dishes. However, such practices also cause weariness, sickness, and pain generated by the preparation process and the precarious social and material conditions in which they work. The following sections will analyse these considerations through the preparation of *Colada de Uchu Jaku*, focusing on the fluctuating experiences of temperature.

Controlling the temperature: heat in preparing, transporting, and selling the soup

Paquiestancia sits between 2,800 and 4,100 metres above sea level, with its population centre based on the slopes of the imposing Cayambe volcano. Eucalyptus forests, primary rainforest, and moorlands surround the village of 1,500 inhabitants. *Doña* Luisa lives with her husband and grandchildren at one side of Paquiestancia's main road. At her farm, *Doña* Luisa grows varieties of potatoes, onions, beans, and corn; broccoli; beets; radishes; chillies; and aromatic herbs like cilantro, parsley, and lovage. She also raises *cuyes* and keeps a few cows for milk and cheese. Her granddaughter Alicia helps her cook; Alicia is a full-time teacher but helps her on these days due to the heavy workload involved.

We stayed that night in a nearby guesthouse on their property. We awoke at 3:00am when the temperature was very low in the skirts of the Andes and walked through the mist to the kitchen, located in a hut. On entering the kitchen, the change in temperature was immediately evident – a blast of hot air hit us like it could have knocked us over. Once we acclimatised, the space felt cosy. We breathed thick smoke from the *tulpa* (a traditional wood-burning stove), where a large amount of water was boiling in a pot. Smoke was everywhere – the walls of the hut were covered with soot from cooking. It got in our eyes and noses quickly – so much that it drew tears. *Doña* Luisa's choice to use a *tulpa* was determined by flavour; the wood gives a particular taste to food that a gas stove fails to replicate. It is in this way that *Doña* Luisa learned to prepare the soup, and how she, in turn, is teaching her granddaughter. The constant smoke that afflicts the eyes and nose is not separate from the emotional fulfilment of her memories. *Doña* Luisa and Alicia appeared accustomed to the smoke and did not show signs of being bothered by it. Furthermore, the *tulpa* warmed up the house while they were cooking, making the labour-intense task more comfortable. The warm room against the cold Andean weather resonated with the warmth we felt in our bodies when consuming the *Colada* later at the *feria*. This was the moment we realised that temperature played multiple, fundamental roles: from the cold weather in the Andean town to the intense warmth generated by the *tulpa*.

Doña Luisa worked with Alicia to prepare the ingredients, adding them to the *Colada*. *Doña* Luisa's children, however, opposed her continuing this laborious process; she has frequent back pain, and her blood levels are not the best, according to her doctor. *Doña* Luisa told us about cooking using the *tulpa* against her doctor's explicit advice, despite suffering consequences in her skin, nails and eyesight. Indeed, domestic burning of biomass and the air pollution it produces are leading environmental contributors to disease worldwide, with women and children particularly at risk (Kim *et al.* 2011; WHO 2002; Rinne *et al.* 2006).

These numerous experiences associated with changing temperatures (Korsmeyer and Sutton 2011; Sutton 2000) are also conveyed by the granddaughter. Alicia talked about how pain accompanies different stages of preparation. The care involved in preparing the soup involves multiple activities that are paired with sensorial experiences. For example, when preparing the *cuyes*, the burning wood or hot frying pan slightly burns her hands, leaving her with pain throughout the day.

Alicia values the family legacy shared by her grandmother. As they placed the meat in the huge pot of boiling water on the firewood, *Doña* Luisa told us how she had learned to prepare the *Colada* from her mother by watching and eventually helping her. She mentioned that until a few years ago, it was not common to find the *Colada* in the city like you find now in the *feria*; they had only made the soup for consumption within rural families or for the community during festivities. She explained how rural families prepared the dish with the crops and animals they had:

> We washed the quinoa, and she toasted it, every single ingredient was carefully prepared, we knew how to make the *uchu jaku* flour. We ate the *Colada* made by ourselves for ourselves. . . . Not to sell. Now I started to sell, before I have never sold. Just for the family.

The *Colada* is a meeting point between generations and a skilful practice that *Doña* Luisa and Alicia share: from peeling the *cuyes* the night before to preparing and placing the ingredients at the appropriate moments, knowing that the *Colada* must be ready at sunrise so they can take it to the market. Through the preparation of the *Colada*, they take care of ancestral knowledge and practices, while new generations have the opportunity to learn and adapt it to the context in which they live. *Doña* Luisa, Alicia, and other producers from the *feria* have adapted a dish previously only for festivities to become a dish offered weekly to diners in the city. The elaborate dish was previously associated with domestic and rural festivities. However, thanks to the labour of the producers in the *feria*, it has become an ordinary weekly urban meal that is available at an inexpensive rate.

Back to the kitchen. Once the lamb-broth base is ready after boiling away, they add the *Uchu Jaku* flour and carefully observe the broth to make sure the flour is cooked through and at the desired consistency. To do this, they have to boil the flour intensively at high temperature, but as the liquid evaporates, it causes the *Colada* to dry, so they add water to keep the soup from getting too thick. At the same time, there is the risk of adding too much water and dissolving the distinctive flavour, so they need to remain vigilant to sense the aromas, textures, and flavours that characterise a well-made *Colada de Uchu Jaku*. *Doña* Luisa´s expertise makes a big difference in the final product. She attends to the art of keeping the right amount of wood burning – at a constant temperature, high but not too great – by adding just enough water so the *Colada* does not get too thick. It is fundamental that *Doña* Luisa tastes, smells, feels, and watches the soup during this stage, which is one of the most exhausting steps of preparation as she has to be near the steaming pot constantly. During this process, the warmth in the hut drops and becomes unpleasant as more water is constantly brought inside to correct the flavour. At this stage, other family members become involved, carrying buckets of water that are too heavy for *Doña* Elisa and Alicia. The texture of the soup changes slowly from a liquid broth to a thicker, bubbly texture with a penetrating aroma of toasted flour. Maintaining the right temperature is crucial to reaching the desired experiences that tell them the soup is ready, but the heat also becomes fatiguing as the *Colada* is reaching this ideal point.

Once the soup is ready, they pack it in metallic milk containers to transport it to the *feria*, making sure that it arrives hot because if it cools down, it will lose the proper texture and thicken too much. Additionally, diners want the soup to be warm and ready to eat. We helped them load a neighbour's truck with the stainless steel twenty-litre container full of soup. They also loaded chairs, tables, homemade hot sauce, cooking utensils, plates, and silverware. The truck then made its way five kilometres down the mountain with its precious cargo and cooks, who had changed their clothes to transform into street vendors. It was around 7:00am by then.

In the *feria*, chaos from the rain made the organisation of food stalls complicated, and customers were waiting. After a quick 'good morning' to fellow producers, the women started assembling missing ingredients, such as the cheese, buying them from other producers and assembling the meal. They placed the *cuy*, egg, cheese, and hominy corn on each plate.

Doña Luisa and Alicia sell each plate for US$3.00, an affordable price for customers who are mainly locals. Some days, as early as 9:30am, the *Colada* is gone. These are recurrent diners, with some taking servings home for their families. The affordable prices in the *feria* also reflect the targeted public. Customers of the *feria* are local working-class people, differing from typically upper-middle-class customers who visit organic markets in large European and North American cities (Anguelovski 2015).

The *Colada de Uchu Jaku* is not cost effective due to the labour and ingredients that are necessary to make it artisanally. For this reason, many members of Bio-Vida prefer to renounce the extra income rather than prepare hot food. For *Doña* Luisa and Alicia, the soup is an option because they grow many of the ingredients it needs. For example, *Doña* Luisa has her own *cuyera* (*cuy* breeding house) to source *cuyes* – one of the most expensive elements of the *Colada.*

Figure 15.1 Colada de Uchu Jaku

Source: Photograph by J. Guillermo Gómez-Urrego

A vital agreement at Bio-Vida is that all ingredients must be either produced by or bought from producers of the *feria*. When *Doña* Luisa does not have ingredients from her plot, she buys them from other members. This condition of exchange creates a very different dynamic to non-agroecological markets because the first and almost only consideration for choosing a supplier is that the person should be part of the organisation. This market is about not only individual profit but also collective support – a particularly crucial criterion when agroecological initiatives often remain overlooked by most consumers in Ecuador.

The *Colada de Uchu Jaku* illustrates the work and time that carers invest to bring different dishes to the table every Wednesday. Although every plate of food has its own long story of labour involving an assemblage of factors, Bio-Vida dishes are unusual in that producers are in charge of almost every single step of the dish. These steps imply a heavy workload, divided into sowing, composting, maintenance, harvest, and preservation, representing complex techniques carried out by different producers in different spaces. Alicia describes this temporal extension and complexity that starts long before the products arrive in the kitchen:

> If you want to see everything, you would have to see it from the beginning when each ingredient is cultivated. You see, just for the flour, you would have to plant corn, beans, wheat, peas, lentils . . . and then would have to wait until the grain comes out. It takes time; it cannot be ready instantly. You have to wait June, July, until August when it is harvested. It is a long process until that point. You need to have all the grains dried. Once you harvest them, you have to dry them, because if you don't do that correctly, they will rot. Then you can start to toast them; grandma has a toasting clay pan, in it, she toasts them.

This quote illustrates only some of the work and care involved in production. Moreover, while the preparation of the dish supports agroecological principles, this labour of care remains invisible for the *feria*'s customers. In light of this, we argue that recognition needs to be dedicated to traditional recipes that have been shared and reproduced through specific bodies and practices under specific circumstances. As long as the labour of care remains unacknowledged inside and outside these alternative models of production, the workload and responsibility will continue to weigh heavily on rural women's shoulders. This means acknowledging painful and exhausting aspects of care alongside the nourishing and satisfying experiences of their labour.

Discussion

We have proposed temperature as an entry point to analyse multiple sensorial experiences as demonstrated through the heating of a specific space (*Doña* Luisa's kitchen), the means to generate that heat (the *tulpa*), how these elements interact with bodies over time (disease, embodied memory, perceived taste and smell), and how warmth travels between rural and urban spaces without ruining the dish (using the milk containers to transport the soup). If we fail to relate food production to the embodied experiences of the carers, we fail to understand their vital role in maintaining healthy bodies and ecosystems. There is a constant problem of possibly erasing labour by only approaching dishes and traditions as public, ephemeral national patrimonies, instead of understanding their connection to place as both joyful and painful experience, an experience that continues to be overlooked.

The story of the *Colada* illustrates that we combine ingredients when cooking that are not essentially paired. People around the world shape their cultural identity by pairing and elaborating ingredients that are significant to them, and therefore cooking becomes a way to relate to their territory in a meaningful way. People cook with and become with the different species that inhabit their different territories, and much care work to sustain the lands is done by women whilst feeding their communities and families. The preparation of the *Colada de Uchu Jaku* by *Doña* Luisa and Alicia illustrates how they are reproducing an ancestral dish rooted in Cayambe and their family's story. They use local products that have been part of Andean culture for centuries, such as the *cuy* and corn. Sustaining this dish also sustains the knowledge that has been shared across generations, combining ingredients that showcase the products of a fertile Andean soil with crops to connect people to their territory in a particularly meaningful way.

Some questions about the sensorial exploration of cooking, food, and the politics of care remain. First, it is important to think in ways that incorporate less-visible experiences and stories of pleasure and pain within the experience of the people visiting the *feria* in Cayambe. This connection across the senses and labour could help support a more conscious and political form of consumer engagement with the *feria* and agroecology more generally. Moreover, it is crucial to make the labour of the carers more visible to people who have the power to mobilise resources and policies, such as local and national authorities, to justly support producers like *Doña* Luisa. Likewise, an important question raised by this story is the physical toll incurred to transform a celebratory dish – previously prepared for special occasions and by a larger group of women – into a commercial dish on the weekly menu of a market in the city. Not to say that there is anything intrinsically bad in commercialising the dish, but rather that its commercialisation implies a heavy workload and also creativity in its adaptation to the rhythms and rules of the city, which are not fully taken into consideration. Finally, the ritual sense of a celebratory dish like the *Colada* – pairing uncommon ingredients, such as different proteins – seems to get lost in its translation into an urban-market context. Traditionally, in the rural areas, it connotes a special temporality when people gather around the preparation of a time-intensive dish. Thus, its connection with the labour embodied within the dish is more evident. Our key argument remains that by exploring the sensorial experiences of carers, we can begin a conversation to mobilise care towards the producers and their work (Puig de la Bellacasa 2017) instead of only valuing their products, be they organic foods or traditional dishes.

Conclusion

The soup is a meeting point between generations with contextual and structural difficulties that involve 'taking the time' to cook the *Colada*, controlling its temperature, and making sure it arrives warm to the diners. Observing the preparation of the dish enabled us to grasp some of its meaning at the *feria*. Our aim was to explore sensorial experiences of food beyond flavours and aroma to pay attention to the experiences of discomfort and pain. We found the incorporation of the *Colada* into the *feria*'s menu illustrates how agroecology connects with traditional knowledge to support a different model of production in which people can feed their families in their territories while supporting healthy ecosystems. However, feeding is still widely gendered labour of care that continues to be undervalued, and infrastructural support remains limited due to the continuous consolidation of agro-industrial production and the broadly uncontested consumption of its products. On a positive note, we hope to see beneficial changes in Cayambe for agroecological producers with the approval of the new law that regulates the use of public spaces. This is a great opportunity to recognise the labour of rural workers who are feeding the city and are maintaining traditions that connect people to their territories in a meaningful, caring way.

References

Anguelovski I. (2015) 'Healthy Food Stores, Greenlining and Food Gentrification: Contesting New Forms of Privilege, Displacement and Locally Unwanted Land Uses in Racially Mixed Neighborhoods', *International Journal of Urban and Regional Research* 39 (6): 1209–1230.

CARE Ecuador (ed) (2016) *Mujeres: Su rol en la soberanía y seguridad alimentarias*. Quito: Ediciones Ciespal, https://drive.google.com/file/d/0B6wJrsxILA__emtsbUw4M21MblE/view.

Centro Peruano de Estudios Sociales. (2011) *Mujer Rural: Cambios y Persistencias en América Latina. Seminario Internacional Mujer Rural: Cambios y Persistencias en América Latina*, www.americalatina.landcoalition.org/sites/default/files/libro Mujer Rural.pdf.

Daza E. (2018) *Razón de estado: cuestión agraria y campesinado*. Quito, Ecuador: OBSERVATORIO DEL CAMBIO RURAL-OCARU e Instituto de Estudios Ecuatorianos.

De la Cruz C. (2018) 'Fundamentos teóricos de agroecología y desarrollo sostenible', in *Curso Virtual. Agroecología Para El Desarrollo Sostenible En Los Andes*. La Paz: IPDRS – Interaprendizaje. Campo virtual por la tierra y el Desarrollo rural.

Duffy M. (2007) 'Doing the Dirty Work: Gender, Race, and Reproductive Labor in Historical Perspective', *Gender and Society* 21 (3): 313–336.

Federici S. (2004) *Caliban y la bruja. Mujeres, cuerpo y acumulación primitiva*. Madrid: Traficantes de sueños.

Federici S. (2018) *El patriarcado del salario*. Madrid: Traficantes de Sueños.

Gortaire R. (2017) 'Agroecología en el Ecuador: Proceso histórico, logros, y desafíos', *Antropología Cuadernos de Investigación* 17: 12–38.

Grey S. and Newman L. (2018) 'Beyond Culinary Colonialism: Indigenous Food Sovereignty, Liberal Multiculturalism, and the Control of Gastronomic Capital', *Agriculture and Human Values* 35 (3): 717–730.

Grey S. and Patel R. (2015) 'Food Sovereignty as Decolonization: Some Contributions from Indigenous Movements to Food System and Development Politics', *Agriculture and Human Values* 32 (3): 431–444.

Hill Collins P. (2000) *Black Feminist Thought: Knowledge, Consciousness, and the Politics of Empowerment* (2nd ed.). New York: Routledge.

Hill Collins P. (2007) 'It's All in the Family: Intersections of Gender, Race, and Nation', *Hypatia: A Journal of Feminist Philosophy* 13 (3): 62–82.

INEC. (2012) *Encuesta De Uso Del Tiempo*, http://www.ecuadorencifras.gob.ec//documentos/web-inec/Uso_Tiempo/Presentacion_ Principales_Resultados.pdf (Accesed: 15 June 2018).

Kim K.-H., Jahan S. A. and Kabir E. (2011) 'A Review of Diseases Associated with Household Air Pollution Due to the Use of Biomass Fuels', *Journal of Hazardous Materials* 192: 425–431.

Korsmeyer C. and Sutton D. (2011) 'The Sensory Experience of Food', *Food, Culture and Society* 14 (4): 461–475.

Larrea C. A. and Greene N. (2018) 'Concentration of Assets and Poverty Reduction in Post-Neoliberal Ecuador', in North L. and Clark T. (eds) *Dominant Elites in Latin America*. Cham: Springer: 93–118.

Macaroff A. (2018) *La Vía Terrateniente: Línea Histórica de las Élites Agrarias en la Política Ecuatoriana*. Quito: Instituto de Estudios Ecuatorianos.

Macaroff A. (2019) 'Oligarquías renovadas. Los grupos de poder en el Ecuador y su accionar político ante los gobiernos progresistas, en el marco del capitalismo transnacional', in Gabbert K. and Lang M. (eds) *Cómo se sostiene la vida en América Latina? Feminismos y re-existencias en tiempos de oscuridad*. Quito: Fundación Rosa Luxemburg-Ediciones Abya Yala: 287–323.

Minga N. (2014) *La agroecología está presente: mapeo de productores agroecológicos y del estado de la agroecología en la sierra y costa ecuatoriana*. Quito: Ministerio de Agricultura, Ganadería, Acuacultura y Pesca – Heifer Ecuador.

Mol A. (2002) *The Body Multiple Ontology in Medical Practice*. Durham, NC: Duke University Press.

Murphy M. (2015) 'Unsettling Care: Troubling Transnational Itineraries of Care in Feminist Health Practices', *Social Studies of Science* 45 (5): 717–737.

Ortega L. (2012) *Las relaciones de genero entre la población rural del Ecuador, Guatemala y México*. Santiago: CEPAL.

Puig de la Bellacasa M. (2017) *Matters of Care: Speculative Ethics in More Than Human Worlds*. Minneapolis: University of Minnesota Press.

Raghuram P. (2016) 'Locating Care Ethics Beyond the Global North', *Acme* 15 (3): 511–513.

Rinne S., Rodas E., Bender B., Rinne M., Simpson J., Galer-Unti R. and Glickman L. (2006) 'Relationship of Pulmonary Function Among Women and Children to Indoor Air Pollution from Biomass Use in Rural Ecuador', *Respiratory Medicine* 100 (7): 1208–1215.

Roberts D. E. (1997) 'Spiritual and Menial Housework', *Yale Journal of Law and Feminism* 51 (9): 51–80.

Salazar Parreñas R. (2015) *Servants of Globalisation Migration and Domestic Work* (2nd ed). Stanford, CA: Stanford University Press.

Sherwood S. G. and Paredes M. (2014) 'Dynamics of Perpetuation: The Politics of Keeping Highly Toxic Pesticides on the Market in Ecuador', *Nature and Culture* 9 (1): 21–44.

Siliprandi E. (2015) *Mulheres e agroecologia: transformando o campo, as florestas e as pessoas*. Rio de Janeiro: Editora UFRJ.

Sutton D. (2000) 'Whole Foods: Revitalization Through Everyday Synesthetic Experience', *Anthropology and Humanism* 25 (2): 120–130.

Tronto J. C. (1993) *Moral Boundaries: A Political Argument for an Ethic of Care*. London: Psychology Press.

Tronto J. C. (2010) 'Creating Caring Institutions: Politics, Plurality, and Purpose', *Ethics and Social Welfare* 4 (2): 158–171.

Whyte K. P. (2016) 'Indigenous Food Sovereignty, Renewal and U.S. Settler Colonialism', in *The Routledge Handbook of Food Ethics* (Vol. 2016). New York: Routledge: 354–365.

World Health Organization. (2002) *World Health Report 2002: World Health Report: Reducing Risks to Health Noncommunicable Diseases*. Geneva: World Health Organization.

16 Future directions for food, senses, and the city

Ferne Edwards

This book sought to carve out a new space to bring together food, senses, and the city. The authors bring primary research from cities across the world to experience – to touch, taste, smell, see, and hear – a range of urban cuisines and to consider their impact on contemporary culture. This final chapter explores the contribution of these narratives and asks: What do the chapters contribute to the nexus of food, senses, and the city? What research topics can further support this trajectory? What is the contribution of this volume to urban food studies literature? To answer these questions, this chapter first summarises the key concepts and topics from the chapters to draw out possible research directions, and ends by summarising the volume's contributions to the wider literature.

Part 1: the city and its other

The first four chapters provide a base on which to build connections through the senses to food and place. Vincent Walstra (Chapter 2) describes how city folk are (re)establishing their understanding of food and nature through urban gardening at Koningshof in the Netherlands. Walstra draws the reader through a history of the deepening alienation of city from country, and of food production from culture, which is being slowly mended through the gentle caring for livestock and tending of soil. The personal stories of urban farmers, complemented by Walstra's own experience, are a reawakening of sorts, in which urban gardeners unpick assumptions about producing one's own food. Walstra finds that 'doing' gardening asserts an appreciation of 'the process rather than the completion of work' (Walstra, this volume), when gardeners go beyond an egocentric idea of resource extraction to share produce with other urban natures. Enskilment is a key aspect through which people (re)learn traditional gardening knowledge and skills through crafted conversations, practice, and patience.

Carole Counihan (Chapter 3) takes this embodiment of food production a step further to consider its role in activism. Counihan combines the 'foundational senses' (Wagenfeld 2009: 48) with Karl Marx's concept of labour (Marx and Engels 1970: 121) into 'sensuous human activity' enabling 'urban consumers to connect holistically with farmers, the land, and food production' (Counihan, this volume). In her chapter, the reader journeys to farmers' fields and livestock yards and is brought to the shared table to join others to explore traditional Sardinian food connections towards making food more sustainable, just, nutritious, and tasty (Counihan and Siniscalchi 2014). Akin to Walstra, Counihan finds the senses are an essential part of 'learning by doing' as illustrated by the failure of a lesson plan on 'his majesty, the Sardinian pig'. Instead, she found that the students 'wanted to see and touch; they did not want to listen' (Counihan, this volume); where a revised approach to a farm visit to see, touch, and listen to the tractor; walk the farm; and finally ingest the pork 'brought home the bacon' to the students, as it were. Other junctures of engagement described by Counihan include a food walk and a visit to an urban garden that reminded the reader of both the artisanry and the sheer hard work that farming requires. Counihan acknowledges that, while food production often plays a small part in activists' lives, the senses serve to 'tap into a need' to remind them both 'why' and for 'what' they are fighting.

Ferne Edwards in Chapter 4 expands the senses to human/nonhuman relations in her study of urban beekeepers in Australian cities. She employs Hayes-Conroy and Hayes-Conroy's (2010) frame of visceral geographies to demonstrate how the senses 'matter' in urban beekeeping. She emphasises how the senses guide reciprocal cross-species' understanding and communication, where mentorship is given as an example of experiential learning, uniting people from different backgrounds to exchange knowledge. Edwards argues that the senses present both a tool and a pathway towards creating 'a multispecies city', where the rights of humans and nonhumans can both be heard.

Roos Gerritsen (Chapter 5) completes this first section by walking the reader through the sense-scapes of Chennai, India, conjuring up its smells, tastes, sights, and sounds to draw the reader closer to a 'vernacular' city. Gerritsen's chapter 'digs in' to the city using a photo-essay approach to experiment with, and go beyond, the limitations of text. She agrees that such techniques 'render different kinds of ethnographic knowledge' (Grimshaw 2005: 18) that pay 'specific attention not only to the sociality that informs the food walks, but also to the food walks as an activity that is both placed and sensed' (Gerritsen, this volume). Place becomes saturated with personal interpretations of the senses, drawing in participants who create their own worlds. This personal and sensual ascription to the city 'is not a backdrop' (Gerritsen, this volume) but instead 'attracts bodies and things; it is part of the way in which subjectivities are formed and formulated' (ibid). The city 'becomes' as people walk along, sense, buy, and consume food within it.

Indeed, Gerritsen's approach evokes the concept of 'the city and the city' penned by science fiction writer China Miéville (2009), in which a myriad of coexisting worlds are revealed and drawn out through the senses. Such multiplicitous versions of the city change over the course of the day, are interpreted differently in each cultural enclave, and are transformed as they move from one edible corner to the next. While each chapter employs a different approach, this iterative, processual, and reflective understanding of 'cities within cities' is representative of many chapters throughout the book.

Part II: past in the present: memory and food

Part II describes how the past is recreated in the present through frames of cultural and culinary nostalgia (Swislocki 2008). Each chapter demonstrates how the senses are not easily dislodged, retaining a persistent power that endures time, place, and cultural change. Homemaking through the senses is explored in two variations: from the physical movement of food over geographical and cultural territory to be reinstated somewhere new, as illustrated by Chapters 6 (Monterescu and Hart), 9 (Bhattacharya), and 10 (van Ommen); to places where the authors, as natives who have once departed, return to rediscover 'home' (Chapters 7 by Wesser and 8 by Chen).

Joel R. Hart and Daniel Monterescu (Chapter 6) analyse nostalgia through the changing perception of *amba*, 'the Iraqi version of mango chutney' (Berg 2019: 77). They trace its journey with Iraqi Jews to London and Israel. Hart and Monterescu remind us that nostalgia is a combination of the Greek *nostos* (to return home) and *algia* (pain), in which restorative nostalgia 'attempts a transhistorical reconstitution of a lost home' while reflective nostalgia 'delays the homecoming, "wistfully, ironically, desperately" (Boym 2001: xviii)'. In Chapters 6, 9, and 10, cultural and culinary diasporas connect history, religion, and politics to taste, as *amba* travels from special prominence in Shabbat culinary rituals to transform into a popular street food today; this process is called the 'falafelisation of the *sabih*' by Hart and Monterescu (this volume). Once outside the Iraqi community, *amba* is reshaped once more to become 'part of the gourmetisation of contemporary Israeli and Middle Eastern cuisine within global cities' (ibid). With its distinctive odour that seeps through the skin, *amba* is hard to ignore. For Iraqi Jews, *amba* represents a lingering connection to a past they can never return to; for others, it becomes a vessel for gastro-racism, conjuring visceral shame (echoing Manalansan 2006), while for still others, it symbolises a celebration of flavour that embraces new beginnings. Drawing on different sources and experiences, Hart and Monterescu find that *amba*'s passage through place articulates 'personal biographies, collective memory, and legacies of cultural syncretism' (Hart and Monterescu, this volume).

Alternatively, Diti Bhattacharya (Chapter 9) examines the confluence of Bengali cuisine that at first seems 'out of place' in the Gold Coast, Australia. Two housemates – one from Dhaka, Bangladesh, the other from Kolkata, India – 'source, sense, and share' each other's home-cooking to create a new sense of belonging in a mutually foreign place. Throughout the chapter, the reader is treated to an assortment of similar ingredients with varied names that reflect their cultural histories. This interchange between cultural cuisines leads the reader through a trajectory of Bengali political history that is accompanied by personal stories in which, in Bhattacharya's words, 'We both adapted to tastes and smells that belonged to Dhaka and Kolkata' (Bhattacharya, this volume). Questions of authenticity and identity come to the fore as the housemates learn to '[mediate] our new identities as complex assembled Bengali migrants' (ibid).

Similarly, Premila van Ommen (Chapter 10) examines the Nepali diaspora in the UK through food images in digital media. van Ommen focuses on networks and systems of media usage applying a polymedia approach that 'takes into account the multiple ways individuals shift between forms of technology according to strategies based on degrees of emotional investment' (van Ommen, this volume). She explains how, following World War Two, 'Gurkhas', an elite military brigade, were absorbed into the police and army forces in former British posts that later grew into distinct cultural groups, many of which relocated to Britain. Increased popularity of curry restaurants and Himalayan street foods in the UK has since led to the publishing of images of Nepali cuisine on social media sites, capturing this cultural diaspora through food images online. van Ommen's research reveals not only how identity construction is formed along the geographical routes of the Gurkhas, but also how 'senses of space and distance collapse' (ibid) as new connections are made virtually with East Asian popular culture and through communicating consumption back home in Nepal.

Grit Wesser (Chapter 7) and Shuhua Chen (Chapter 8) present the viewpoint of anthropologists returning home, where they reconnect with their memories through re-enchanting with local foods. Wesser's account concerns the importance of cake in Thuringian households, where cake expresses, amongst other things, familial love. In Wesser's words: 'my mother expressed her love through homemade cake and that my being with them for a year was a kind of special – almost unimaginable – family reunion' (Wesser, this volume). The joy of baking is captured through Wesser's historical account of *Kaffee und Kuchen* (coffee and cake), a Sunday ritual established in West Germany in the 1950s that associates eating cake with a harmonious family home. On their first visit to West Germany after the fall of the Berlin Wall, East Germans were struck by a sensory overload, and Thuringian cakes offered a way to recuperate a simpler sense-scape to 'make Thuringians feel at home in Thuringia and, ultimately, in contemporary Germany' (Wesser, this volume). Kinship, gender, and commensality emerge as key themes within this postsocialist narrative, where Wesser shows the importance of the emotional and sensuous processes in belonging that – concurring with Janet Carsten (2004) and Tatjana Thelen and Erdmute Alber (2018: 12–13) – can be 'scaled up from familial to regional home – with the potential to incorporate the nation', as evidenced by Thuringian festive cakes.

Shuhua Chen (Chapter 8) examines rural-to-urban migration in China, focusing on migrant workers' experiences of urban marketplaces. Also employing a photo-essay approach, Chen walks through a local fresh food market in conversation with her interlocutor, Yang Cui. This 'conversational' food walk captures both emic and etic perspectives at varying degrees of temporal proximity, as Chen, a native of the region, knows the location intimately but not recently while her companion is a stranger to the area yet frequents its market rows often. Chen describes how the market in Shantou is 'scaped', both in the process of '*being scaped* (shaped) and the *scape* (shape) itself', where both scapes entwine to produce embodied experiences that are continually (re)shaped (Chen, this volume). Through this sensual foodscape, Chen 'was able to inhabit Shantou's manufacturing landscape, breathe air that was polluted due to unchecked industrialisation, and taste its evolving foodscapes' (ibid), whilst drawing on her own familiarity with the site entangled with Yang Cui's perspective. This narrative introduces cultural concepts such as '*mianzi*' ('face' in Chinese) - *you-mianzi* (having face) and *mei-mianzi* (having no face) (André 2013) - that are ascribed through the social judgements of others participating in local shopping practices. So, too, does local folklore live on, where this account links the Year of the Dragon with a scarcity of pig kidneys and hearts, a consequence of '*yixingzhixing*': eating an animal part to strengthen the corresponding part of the human body. Chen demonstrates how 'tastes of home' are ascribed differently through memory and place as her and Yang Cui's interpretations differ across the foodscape. Yang Cui, feeling more at home in the part of the market that sells familiar foods, is able to 'practice' home, even momentarily (Chen, this volume) in her brief morning visits to the market. Further, Chen acknowledges how, as they move through the space of the market, 'we were *re-scaping* our sense of belonging' (Chen, this volume), creating new dimensions of being at home and away, entwining past with the present.

Part III: disrupting and re-imagining

The final section further shatters static conceptualisations of a 'normal' city or a 'one-for-all' cuisine to recognise how cities and their diets are changing. Melissa S. Biggs (Chapter 11) opens the conversation by introducing the notion of the *paladar tapatío* ('typical plate') to ask, 'What does a city taste like?' This question prompts a discourse on plurality, agency, and the confluence of taste, as food and city metamorphose into new states. However, in what direction they move is another matter entirely. This section offers three possibilities: towards gentrification and homogenisation through external pressures; through a confluence of factors across the city and from rural/urban linkages; and through justice and sustainability fostered by internal desires, such as activism.

Chapters 11 (Biggs), 12 (Stroe), and 13 (Leizaola) explore this first possibility where urban food practices encounter pressures of gentrification. Biggs takes us to Guadalajara, Mexico, where she sets off to explore:

> the ways Guadalajara residents experience and talk about food to better understand what residents found unique and delicious about their city's food and what they wanted visitors to know about it.
>
> (Biggs, this volume)

Very soon, we find that traditional plates are being impacted by a mix of factors that include the conversion of buildings into condominiums, the arrival of co-working spaces and Walmart and the restructuring of the local market (to become cleaner, tidier, and faster), and an 'influx of sugar and industrialised corn- and soy-based snacks' that introduce new tastes (Biggs, this volume). Similarly, Monica Stroe (Chapter 12) examines the appropriation of the traditional working class snack of *mici*, a Romanian meatball, that is historically linked to Labour Day celebrations. Stroe's focus is not on the people who ordinarily consume *mici*, but on those who seek it out as 'eatertainment'. Such 'foodie *flaneurs*' represent a postsocialist middle class that 'mirrors global gastronomic and nutritional trends' aiming to attain culinary capital fuelled by 'an individual's openness to a range of experiences (Naccarato and LeBesco 2012: 9)'. Stroe contends that such 'foodies erode the logic of class-based divisions of taste' to instead privilege authenticity and exoticism (see Johnston and Baumann 2014).

Aitzpea Leizaola (Chapter 13) investigates the appropriation of the *pintxo* in Donostia, Basque Country. As in Stroe's and Biggs' chapters, a material, symbolic, and cultural distancing occurs between traditional and new forms of consumption. However, for Leizaola, this transformation occurs with respect to the arrival of foreign foodies who travel to Donostia to appreciate its cultural heritage. New forms of material culture (plates, cutlery, and chairs) are introduced to accommodate new consumers who change the use of urban (indoor and outdoor) spaces and eating behaviours. With an almost-complete change of audience in local bars, the consumption of their traditional dish is transformed, resulting in locals no longer feeling welcome in their own city. Such a shift in the performance of a specific cuisine demonstrates the prominent and powerful relationship of food to heritage, memory, and place, prompting questions of inclusion, homemaking, and whether cities can accommodate all people and their cuisines.

Catherine Earl (Chapter 14) examines rural/urban interactions with respect to their meals in Ho Chi Minh City, Vietnam. She demonstrates how 'culinary experiences may be transported between locations and shared with relatives and others in different places' (Earl, this volume). Through her primary research, we discover how Vietnamese sense-worlds are heterogeneous and diverse, characterised as 'indifferentiation' by Lahire (2008), which, in Vietnam 'smooths over assumed dichotomies of low-brow and high-brow, rural and urban, traditional and modern tastes' (ibid). Here, urban food practices move across the city, adapting to new locations, carrying with them personal stories that, in turn, shape the evolution of the city. Such changes move across the city and extend beyond its immediate boundaries, expanding the conceptualisation of 'urban' food. Earl reminds the reader of the importance of urban/rural linkages that tend to be largely ignored by urban food scholars yet introduce salient factors of growing cities, globalisation, population growth, and resource management (see FAO 2019).

Finally, Paz Saavedra *et al.* (Chapter 15) provide an example of grassroots change, in which advocates of agroecology literally fight for a space in the city to sell and promote agroecological produce. They foreground this fight with the story of *Doña* Luisa and her granddaughter Alicia as they endure extremes of temperature and other physical exertions to prepare the traditional dish, '*Uchu Jaku*' for the weekly market. Saavedra *et al.* apply a framework of care to examine this rural/urban activity, in which the authors begin their research in the protagonist's home in the hills to prepare the dish to travel with them to the market. Care is practiced holistically in agroecology (Mol 2002); 'producers maintain ancestral knowledge by providing nutritious meals to their families while connecting consumers at the *feria* to the principles and values of agroecology' (Saavedra *et al.*, this volume). However, care can also be painful. Saavedra *et al.* (this volume) argue:

> If we fail to relate food production to the embodied experiences of the carers, we fail to understand their vital role in maintaining healthy bodies and ecosystems. There is a constant problem of possibly erasing labour by only approaching dishes and traditions as public, ephemeral national patrimonies, instead of understanding their connection to place as both joyful and painful experience, an experience that continues to be overlooked.

These chapters prompt discussions on the possible new directions for cities where the senses in food practices open up new understandings and considerations. Through these diverse narratives, we find stories of place-based diversity, resilience, and resistance against the homogenisation of cuisine, where provocateurs question, negate, and prompt emerging flavours of and for the city. The next section considers possible directions for future research.

Suggestions for future research

This volume provides a glimpse of the range of topics that can be explored within the 'food, senses, and the city' nexus. While chapters in this volume are not representative of all on offer, they highlight common emerging themes – home, class, identity, conviviality, gender, the body, mobility, and ethnicity – whilst emphasising possible research directions. To revisit some themes in more detail, while 'mobility' is no longer the book's explicit focus (see Preface), this theme underpins many chapters in which all authors have iteratively led the reader through a multi-sensory narrative of distinct urban sensory worlds. These fluid, relational, cultural, place-based accounts blur traditional boundaries and divisions to reflect changing cultures, their diets, and their relationship to place. With many chapters specifically focusing on diaspora, it often becomes difficult to pinpoint where a dish is from, ushering in new understandings, such as home and belonging, whilst challenging traditional static conceptualisations, such as authenticity. Rather than consider cultural cuisine as an isolated event, a flow, blend, and exchange across cuisine, cultures, and disciplines becomes apparent. This relational flow across culture, city, food, and the senses has 'significantly deepened our sociological understanding of previously under-theorised aspects of experience' (Rhys-Taylor 2013: 394).

Some themes could be further developed while others are yet to be explored. For example, we note that studies of the senses in indigenous communities are entirely overlooked, where examples such as Kathryn Geurts' (2003) study of the Anlo Ewe in Ghana that expands the senses to consider 'attention' would be highly welcome with respect to food practices. So, too, would research on different cultural perceptions of the senses, food, and place be rewarding. As recognised in Chapter 1, senses are experienced differently in different cultures; this book adds material on how they change on contact across cultures and their transformation as they pass through different mediums. Examples given include Spanish *convivencia*, flavourful postsocialist states, and the blends of Asian cultures undergoing diaspora from China, Vietnam, India, Bangladesh, and Nepal. This nexus could be explored in more detail, both within these regions and in new locales (for example, the North and South Poles, the USA, island nations, and so forth) to better understand how cultural interpretations of the senses impact understanding, desire, and practices of food in the city. Scant attention has been placed on this trifecta of topics, standing to open insightful terrains, as illustrated by the beginnings of chapters here and by work such as that of Hayes-Conroy and Hayes-Conroy (2010), who go beyond culturally bland assumptions to explore how food feels for participants from different cultural backgrounds in the Slow Food movement.

The senses offer new directions for the changing city that call for fresh ways to adapt to urban life. Potential applications include how the senses in urban food practices could provide forms of inclusion and support, sustain, and empower communities in times of urban disaster catalysed by climate change, disease, or political uprisings. Alternatively, the senses could help solidarity efforts to resist unwanted external pressures or, indeed, could help shape urban food practices to purposefully create desired city 'tastes'.

Another area of research to be further developed is the politics of the senses that could emphasise wider structural issues that limit people's ability to sense city spaces, their produce, other species, and each other. Following on from Monica Degen, who argues that space is political, 'meaning that social power relations are expressed in and through space' (Degen 2008: 10), we argue that politicised spaces are also sensory. In order to react to injustice, we need to make sense of what is happening. Food represents a common practice through which people can (re)connect on material, sensorial, and symbolic grounds towards acting on better futures.

We next return to methodology to consider how scholars can capture sensual experiences as they manifest, move and transform, and are shared and ingested. Based mainly in anthropology, the authors in this volume draw on ethnographic research using qualitative methods of interviews and participant observation to provide grounded and holistic accounts in which to emplace the senses. Many authors are also part of the research, so their personal encounters with senses, food, and place add reflexivity.

Some authors employ novel techniques, such as food walks and photo-essays, to capture both moments and mobility, where en route conversations provide thick description to refine where the camera takes its pause (see Chapters 5 and 8). New media technologies are another example that allow for mediated, relational culinary exchange to occur, tempering and spicing 'local' consumption (see Chapters 10 and 12). While much of the book remains based in text as its primary form of communication, different styles of text are used to portray the fluidity, relationality, and multi-perspectives of the case studies. These include 'follow-the-people' (Marcus 1995) and 'follow-the-food' (Cook *et al.* 2004) approaches as attempts to de-privilege tendencies for static places and ocularcentrism that can limit holistic understandings (see Chapters 6 and 14). By no means are we claiming that these approaches are cutting edge or complete – indeed, David Howes (Pink and Howes 2010) in the key debates raised in Chapter 1 deplores Sarah Pink's suggestion that food walks are an example par excellence. Instead, in this volume, we showcased a sampling of approaches to supplement and reinvigorate this conversation in the context of urban food to demonstrate both nuances within current approaches and their possibility for expansion. Further research in both more-than-text methods and alternative approaches within text would benefit this topic. 'Deep mapping' is one such form of representation that could embrace the senses in a 'more-than-text' way. Described as an embodied and reflexive immersion process 'that is lived and performed spatially, "deep mapping" is a cartography of depth. A diving within' (Roberts 2016: 6).

Contribution to urban food studies literature

There is a tendency to believe that we are losing our senses with greater urbanisation, distance from our food sources, and as a reaction against health risk. This book as a whole pushes back from the perception of the growing sameness of cities to illustrate that diverse, sensual experiences remain in a myriad of urban places, not bound by place but existing as mobile and adaptive. The chapters here join other examples in showing how the sensoria play an important role in producing the 'other' in people and places (Rhys-Taylor 2017: 12; see also Manalansan 2006). What becomes even more apparent through this nexus is not examples' similarities but more their differences, illustrating how people, places, and food types remain unique, sustained, and celebrated, adapted yet distinct.

C. Nadia Seremetakis acknowledges that the 'numbing and erasure of sensory realities are crucial moments in socio-cultural transformation on sensory experience, memory and material history' (Seremetakis 1993: 2). In this volume, we have sought to 'make visible' both these articulations of changing urban cultures and the urban diversity that remains. These voices include the dispossessed, the transient, the marginalised, and the hidden. Seeking to give voice to the powerless and underprivileged, *Food, Senses and the City* provides an inclusive frame in which a variety of methods can be embraced to engage, empathise with, and represent a wide spectrum of urban experiences.

This volume has also shown the power of the senses in food practices and their place in memory, identity construction, social groupings, and activism. These narratives reveal many (re)conceptualisations of food and 'the' city: where food entwines with the 'vernacular city'; the 'multispecies city'; a city that extends beyond fixed geographical boundaries; and a city that harbours pain, struggle, contestation, (re)connection, and difference and as a way of finding home.

Returning to the question of interdisciplinarity raised in Chapter 1, a range of disciplines helps inform theory and practice – geography, science and technology studies, and cultural theory – helping the reader better understand conceptualisations of space, place, technology, and mobility. We argue that anthropology contributes a grounded, holistic approach that can benefit from and be a benefit to other disciplines to extend beyond textual representation to capture and convey the senses to a wide audience.

Drawing from anthropological theory and approaches, informed by related disciplines, we aimed to deliver what David Howes (Pink and Howes 2010: 340) decries as a need for:

> 'thickly-textured studies'– engaging with primary material, informed about significant work in the field, and raising new issues for consideration – are also what are most needed in the area of sensory anthropology.

Conclusion

The chapters in this book represent multi-sensory accounts of an emerging 'food, senses, and the city' nexus that is becoming more pronounced as our cities grow larger and more complex, where sensing in urban food practices can help distinguish, enrich, and reconnect people to their food, to place, and to each other. This final chapter highlights key contributions to this nexus that emphasise themes of home, belonging, identity, and memory. Spanning urban centres from Europe to Asia, Australia, and the Americas, this collection occupies new territories for further expansion. In doing so, this volume also prompts further questions: How does your city feel through food? How can we add the subjective as a valid indicator to place-making in urban food projects? How can we show that the senses and food matter? What are the politics in bringing senses into the discussion on food in the city? Are senses in urban food gendered, racial, or classed? We have begun to answer some of these questions in this volume. However, rather than attempt to answer everything within one small volume, we hope that we have conveyed that 'food, senses, and the city' is a valuable, thought-provoking nexus to explore, and we invite others to join us in developing this field.

References

André J. (2013) 'How the Chinese Lost "Face"', *Journal of Pragmatics* 55: 68–85.

Berg N. E. (2019) 'Jews Among Muslims: Culinary Contexts', in Diner H. R. and Cinotto S. (eds) *Global Jewish Foodways: A History*. Lincoln: University of Nebraska Press: 70–90.

Boym S. (2001) *The Future of Nostalgia*. New York: Basic Books.

Carsten J. (2004) *After Kinship*. Cambridge: Cambridge University Press.

Cook I. *et al.* (2004) 'Follow the Thing: Papaya', *Antipode* 36 (4): 642–664.

Counihan C. and Siniscalchi V. (eds) (2014) *Food Activism: Agency, Democracy, Economy*. Oxford: Bloomsbury.

Degen M. (2008) *Sensing Cities: Regenerating Public Life in Barcelona and Manchester*. London, England: Routledge.

Food and Agriculture Organization of the United Nations (2019) *City Region Food Systems Programme. Reinforcing Rural-Urban Linkages for Climate Resilient Food Systems*. Rome: FAO.

Geurts K. (2003) *Culture and the Senses Bodily Ways of Knowing in an African Community*. Berkeley: University of California Press.

Grimshaw A. (2005) 'Eyeing the Field: New Horizons for Visual Anthropology', in Grimshaw A. and Ravetz A. (eds) *Visualizing Anthropology: Experimenting with Image-Based Ethnography*. Portland: Intellect: 17–30.

Hayes-Conroy A. and Hayes-Conroy J. (2010) 'Feeling Slow Food: Visceral Fieldwork and Symmetrical Research Relations in the Alternative Food Movement', *Geoforum* 41 (5): 734–742.

Johnston J. and Baumann S. (2014) *Foodies: Democracy and Distinction in the Gourmet Foodscape*. New York: Routledge.

Lahire B. (2008) 'The Individual and the Mixing of Genres: Cultural Dissonance and Self-Distinction', *Poetics* 36: 166–188.

Manalansan M. F. (2006) 'Immigrant Lives and the Politics of Olfaction', in Drobnick J. (ed) *The Smell Culture Reader*. London and New York: Bloomsbury: 41–52.

Marcus G. (1995) 'Ethnography in/of the World System: The Emergence of Multi-Sited Ethnography', *Annual Review of Anthropology* 24: 95–11.

Marx K. and Engels F. (1970) *The German Ideology*, Arthur C. J. (ed). New York: International Publishers.

Miéville C. (2009) *The City & The City: A Novel*. New York: Random House Reader's Circle.

Mol A. (2002) *The Body Multiple Ontology in Medical Practice*. Durham, NC: Duke University Press.

Naccarato P. and LeBesco K. (2012) *Culinary Capital*. London and New York: Berg.

Pink S. and Howes D. (2010) 'The Future of Sensory Anthropology/the Anthropology of the Senses', *Social Anthropology* 18 (3): 331–340.

Rhys-Taylor A. (2013) 'The Essences of Multiculture: A Sensory Exploration of an Inner-City Street Market', *Identities* 20 (4): 393–406.

Rhys-Taylor A. (2017) *Food and Multiculture: A Sensory Ethnography of East London*. London: Bloomsbury.

Roberts L. (2016) 'Deep Mapping and Spatial Anthropology', *Humanities* 5 (5). DOI: 10.3390/h5010005.

Seremetakis C. N. (1993) 'The Memory of the Senses: Historical Perception, Commensal Exchange and Modernity', *Visual Anthropology Review* 9 (2): 2–18.

Swislocki M. (2008) *Culinary Nostalgia: Regional Food Culture and the Urban Experience in Shanghai*. Stanford, CA: Stanford University Press.

Thelen T. and Alber E. (eds) (2018) *Reconnecting State and Kinship*. Philadelphia: University of Pennsylvania Press.

Wagenfeld A. (2009) 'It's More Than Seeing Green: Exploring the Senses Through Gardening', *Journal of Therapeutic Horticulture* 19: 46–52.

Index

Note: page numbers in *italics* indicate a figure on the corresponding page.

For Product Safety Concerns and Information please contact our EU
representative GPSR@taylorandfrancis.com
Taylor & Francis Verlag GmbH, Kaufingerstraße 24, 80331 München, Germany

www.ingramcontent.com/pod-product-compliance
Ingram Content Group UK Ltd.
Pitfield, Milton Keynes, MK11 3LW, UK
UKHW020148200926
475702UK00016B/2610
* 9 7 8 0 3 6 7 7 2 3 6 2 0 *